Use of the Woodlands in the Late Anglo–Saxon Period

R.D. Berryman

BAR British Series 271
1998

Published in 2019 by
BAR Publishing, Oxford

BAR British Series 271

Use of the woodlands in the Late Anglo-Saxon Period

© R. D. Berryman, 1998

ISBN 9780860549475 paperback
ISBN 9781407319025 e-book

DOI https://doi.org/10.30861/9780860549475

A catalogue record for this book is available from the British Library

This book is available at www.barpublishing.com

BAR Publishing is the trading name of British Archaeological Reports (Oxford) Ltd.
British Archaeological Reports was first incorporated in 1974 to publish the BAR
Series, International and British. In 1992 Hadrian Books Ltd became part of the BAR
group. This volume was originally published by Archaeopress in conjunction with
British Archaeological Reports (Oxford) Ltd / Hadrian Books Ltd, the Series principal
publisher, in 1998. This present volume is published by BAR Publishing, 2019.

BAR
PUBLISHING

BAR titles are available from:

BAR Publishing
122 Banbury Rd, Oxford, OX2 7BP, UK
EMAIL   info@barpublishing.com
PHONE   +44 (0)1865 310431
FAX   +44 (0)1865 316916
www.barpublishing.com

Contents

Photographs

Figures

Abbreviations

Abbreviations have been kept to a minimum. For books, the author, and date of publication are given.

Abbreviations used:

BAR(BS)- British Archaeological Reports,
BAR(SS)- British Archaeological Reports,
B.L.   - British Library
E.E.M.F.   - Early English Manuscripts in
Facsimile
E. E. T. S. - Early English Text Society
E.P.N.S.-   English Place Name Society
M.H.G.   -   Middle High German
O.E.   - Old English
O.H.G   - Old High German
O.N.   - Old Norse

An Anglo-Saxon Dictionary Based on the Manuscript Collections of the Late Joseph Bosworth Edited and enlarged by T. Northcote Toller with revised and Enlarged Addenda by Alistair Campbell Oxford, 1898-1921 is shortened to 'Bosworth and Toller Anglo-Saxon Dictionary'. *The Supplement by T. Northcote Toller with revised and enlarged Addenda by Alistair Campbell.* Oxford 1921, reprinted 1972 is referred to as 'Bosworth and Toller Anglo-Saxon Dictionary (Sup)

Where quotations are given in Old English the character wynn (ƿ) is replaced with the letter **w.**

The manuscript symbol 7 has been replaced with *and* in English and *et* in Latin.

Reference to P.H.Sawyer, *Anglo-Saxon Charters: An Annotated List and Bibliography.* London, 1968, is made by quoting the charter number as (S. + number).

The Linnaean system of binominal nomenclature has been used for plants. The binominal system uses the description of the generic name, such as *Quercus* the genus 'oak', which is followed by the specific name, such as *robur* or *petraea* When the species is unknown the abbreviation 'sp.' is used, or 'spp.' to indicate any of a number of species. One item, made of wood, such as a cup, will be referred to as *Quercus sp.*, as only one (unknown) species is referred to, but oak trees or articles which may be of a variety of species will be *Quercus spp.*

[W.T.Stearn (introduction) to Carl Linnaeus *Species Plantarum*, a facsimile of the first edition of 1753, volume I printed for the Ray Society, London 1957, pp. 84-85.]

Abstract

The central theme of this work is that of the reliance of the Late Anglo-Saxons on woodland.

The management of woods is examined and examples of the use of the coppice crafts of wickerwork and basketry in the making of buildings and containers is looked at. Hurdles, forks, and besoms are illustrated and the method of manufacture described.

The section which describes examples of the use of large trees, 'standards' is selective and looks at pole lathe turning, coopering, and carpentry. The north door of St.Botolph s Church, Hadstock is chosen as an example of fine craftsmanship, with quarter sawn planks and bent wood. The occurrence of wooden buildings is discussed and the techniques to be seen at the wooden church at Greensted are described.

The use of tree parts - bark, twigs and fruit - in the Leechdoms is considered, and the available dyes from leaves, bark and berries are listed.

The use of wood as a fuel is considered, and attention drawn to the large demand. The use of wood ash in glass making, and the production of lye and soap is described and experimental notes are appended. The tanning of leather using lye and oak bark is described.

Charcoal making and its use in smelting is briefly outlined and the by-product of tar examined.

The smelting of metals is discussed and the large amount of fuel required is considered.

Acknowledgments

The search for the information about the Anglo-Saxon industries based on the woodlands has taken my wife and I the length of England. Visits have been made to woodlands and parks in search of boundary banks and deer park walls; to coppices and churches; to places as diverse as the Manchester Police Museum and the glades of Hockley Wood. Help and interest has been freely and enthusiastically given, and I acknowledge help from museum curators and strangers on the far end of telephones. The welcome which we received at Hadstock by Mrs. L. Dawson and Mr. D. Stewart gave us a good start and the wardens at the Groundwork Trust, Park Bridge, Oldham, Rawtenstall and Knowsley encouraged me in the search for traces of Anglo-Saxons. The interest and help from Mr. M. Spindley, head of science at Kaskenmoor Comprehensive School, Oldham enabled the 'Soap Project' to take place with pH calibrations and was a good sounding out place for ideas. Bill Hogarth and Walter Lloyd gave valuable time and information on coppice crafts and charcoal burning.

The help given by the librarians at the John Rylands university library, both on the campus and at Deansgate, has been courteous and efficient. I have received valuable guidance and encouragement from my tutor, Dr.David Hill, who has given me suggestions and contacts in my research.

During my three years of study I have had the good fortune to have colleagues who have shared their interests and skills with me, and contributed to a companionship within our group. The tutors assigned to us have been helpful and friendly. I must mention Dr. Alex Rumble, who has taken a personal interest in us all. I must also acknowledge the patience of friends, such as Harry Bintley and David Fielding who have helped me with the mysteries of computers. Finally I must say thank you to my wife, Janet, for her photographs and line drawings, encouragement and patience while I have wrestled with my word processor.

INTRODUCTION

The study of the Late Anglo-Saxon Period has many facets, and the topic of this work draws together many different crafts and skills on which the Anglo-Saxons depended. Although evidence may be sought in charters and Domesday Book, in drawings from psalters and calendars and archaeological discoveries, it is the interpretation of the accumulated facts which help to build a mental picture. Charters are few, especially when only one topic is being sought or researched. Drawings may be influenced by classical traditions, so that the objects drawn may not be of the period. Domesday Book gives details the import of which may not be immediately recognised. Archaeological finds are dependent on a great many factors. Current legislation enables preliminary investigation on sites which may be of archaeological importance. However chance finds, such as the fish weir at Colwick, depend on workmen on the spot who recognise the importance, and on the contractor being willing to work elsewhere while the excavation and recording are in progress. This of course means that some evidence is unrecognised or deliberately ignored.

Place-name evidence can reinforce or confirm evidence from other sources. Traditional crafts have been included in this study as some, such as basketry, are part of an undoubted continuum. Others, such as soap making, are not so obviously recognisable as having existed in the Late Anglo-Saxon Period. It is the cumulative evidence from different sources, archaeological, documentary and traditional crafts, from within Anglo-Saxon England or its close neighbours in both time and space, which reinforce the argument that a method or practice was used in the Late Anglo-Saxon Period.

The use of woodlands as a resource is referred to by King Alfred in his preface to *St. Augustine's Soliloquies*. Although this is used in a metaphorical sense to introduce a religious philosophical discussion, it is the sense of the usefulness of the woodlands which is relevant here. The importance of choosing something with which his readers would be familiar underlines the essential quality of the woodlands, a reliable resource upon which the comforts of this world depended.

King Alfred's Version of St. Augustine's Soliloquies.
(from) King Alfred's Preface. (Carnicelli, 1969)

> Gaderode me þonne kiglas, and stuþansceaftas, and lohscealtas, and hylfa to ælcum þara tola þe ic mid wircan cuðe, and bohtimbru and bolttimbru to ælcum þara weorca þe ic wyrcan cuðe, þa wlitegostan treowa be þam dele ðe ic aberan meihte. Ne com ic naþr mid anre byrðene ham, þe me ne lyste ealne þane wude ham brengan, gif ic hyne ealne aberan meihte. On ælcum trewo ic geseah

> hwæthwugu þæs þe ic æt ham beþorfte. Forbam ic lære ælcne ðara þe maga si, and manigne wæn habbe, þæt he menig to þam ilcan wudu þar ic ðas stuðansceaftas cearf, fetige hym þar ma, and gefeðrige hys wænas mid fegrum gerdum, þat he mage windan manige smicerne wah, and manig ænlic hus settan and fegerne tun timbrian þara. and þær murge and softe mid mæge on eardian ægðer ge wintras ge sumeras, swa-swa ic nu ne gyt ne dyde.

This translation is the one given by Simon Keynes and Michael Lapidge (1983), which is, in the author's opinion, more satisfactory than that given by Henry Lee Hargrove (1902).

> I then gathered for myself staves and props and tie-shafts, and handles for each of the tools that I knew how to work with, and cross-bars and beams, and, for each of the structures which I knew how to build, the finest timbers I could carry. I never came away with a single load without wishing to bring home the whole of the forest, if I could have carried it all - in every tree I saw something for which I had a need at home. Accordingly, I would advise everyone who is strong and has many wagons to direct his steps to that same forest where I cut these props, and to fetch for himself and to load his wagons with well cut staves, so that he may weave many elegant walls and put up many splendid houses and so build a fine homestead, and there may live pleasantly and in tranquillity both in winter and in summer- as I have not yet done.

This document is M.S. B.L. Cotton Vitellius A XV fol. 4a through 59b, which has been dated to the 2nd quarter of the 12th century. Thomas Carnicelli discusses the problems raised by the technical terms in translation. Kiglas, entered in *An Anglo-Saxon Dictionary (Sup.)* as cycgl, from which our word 'cudgel' is derived (Murray and Bradley, 1888-1928) appears to have changed its application. Lohsceaft created a problem for Bosworth and Toller (1838-1921), where the explanation 'a bolt, bar' is offered with some caution. The word stuþansceaft survives in the building term 'studded wall', but other words are not clear in meaning. The nature of the passage, with alliteration, as with bohtimbru and boltimbru , and the underlying theological allusions implicit in a work of this nature, make a full understanding almost impossible, principally because of nuances and double meanings which are long forgotten.

This work shows how the woodland products were also used to provide baskets and leather goods, iron and soap, fences and medicines. Many other uses of wood, such as fuel for the burning of limestone for plaster and cement, may only be inferred.

Part One
Woodland Management in the Late Anglo-Saxon Period

Woodland management is defined by Oliver Rackham who considers the distinctive features of managed woodland to be that the woods are named, are privately owned, have definite boundaries, and are managed by rotational felling to provide a succession of crops protected by fencing. Mr.W. Hogarth, a coppice worker of Cumbria, explains that when woodland is unmanaged it becomes overgrown and choked so that diversity of wild flowers, birdlife and small mammals becomes limited. One cannot but wonder if this is referred to in the Domesday entry for part of the estate which included Doncaster, Warmsworth, Balby, Loversall, Littleworth, Austerfield and Aukley, Yorkshire,

> Silva per loca pastilis. per loca inutilis.
> Woodland, in places pasturable, in places useless.
> (Margaret L.Faull & Marie Stinton, 1986)

The following evidence for the five features of managed woods shows that Anglo-Saxon woodland was indeed managed. .

(1) Woodland Names.

There are three sources which show that woods were named in the Late Anglo-Saxon Period. Domesday Book records woods which, being established by the date of Domesday (1086), may be considered to have been present in the Late Anglo-Saxon Period.

Few are named in Domesday, but in the Leicestershire entry we read;

> Silva totius Vicecomitatis Hereswode vocata
> Woodland of the whole Sheriffdom called Hereswood.
> (Philip Morgan, 1979)

The Yorkshire entry for Wakefield reads;

> Sup' æcclam S. MariÆ quæ e in silva Morlege
> Concerning St. Mary's Church which is in Morley Wood
> (Faull & Stinton, 1986)

Charter evidence is more abundant, and in the charter (S. 1305) granting the lease of land by Werefrith, Bishop of Worcester, to Cyneswith, we have;
> se Alhmunding snæd here into Preasdabyrig
> Elmstone Wood shall belong to Prestbury
> (A.J.Robertson, 1939)

Asser in his *Life of King Alfred.* refers to Berroc wood as giving its name to Berkshire.

> Quae nominatur Berrocshire; quae paga taliter vocatur a Berroc silva, ubi buxus abundantissime nascitur.

> Known as Berkshire, so named from Berroc Wood, where the box tree grows abundantly.
> (W.H.Stevenson, 1904)

(2) Private ownership.

Ownership of the woodland is indicated in Domesday within the entries for manors listed under the landholder's name or title. In Kent the landholders include King William, the Archbishop of Canterbury, Battle Abbey and Richard of Tonbridge. (Morgan, 1983) Charters also indicate ownership of woodland, as indicated above in the grant of lease by Werefrith to Cyneswith of Elmstone Wood. There is also the granting of rights, which implies ownership by the grantor, as in the charter for Bentley of 866 (S. 212), which grants two *manentes* in Seckley, Wolverly;

> V plaustros plenas de virgis bonis et hunicuique anno unum roborem ad ædificium *et* alias materias necessarias. *et* lignaria exabuntia ad ignem sicut illi necesse sit *et* aliam silvaticam taxadtionem. pertinentam ei etiam dabo altrinsecus in campo *et* silvo sicut ad terram duorum manentium pertinent.

> Five wainloads of good brushwood, one oak annually and other timber necessary for building, firewood sufficient for his needs and other rights in woodland and open land pertaining to two *manentes* (W.de Gray Birch, 1885-1899)

(3) Fences

In Kent, under the entry for Wingham, there is recorded

> ii silv,' ad clausuram,
> two small woods for fencing.'(Morgan 1983)

and in Cambridgeshire, under the entry for Haslingfield, we have

> Nemus ad sepes refic
> Wood for repairing fences
> (Alexander Rumble, (ed), 1981)

Hedges existed in Saxon times, and both W.G.Hoskins and Rackham believe that many still survive. Old English words for hedge, hege, hegeruwe, ræw, are to be found in Bosworth and Toller's *An Anglo-Saxon Dictionary.* and the associated words geat (gate) and stigol (stile) indicate access to the enclosure. The variety of hedges referred to, hedgerows, hazelrows, thornrows, willow-rows and hawthorn suggest that hedges were a familiar sight in Saxon times and were deliberately made of selected varieties of tree. Fields and gardens would also require enclosing with fences and hedges, as well as the woodlands.

(4) Boundaries

The evidence for boundaries is not so apparent but it may be significant that the extent of the woods is defined in Domesday by length and breadth, while meadows are recorded by acreage and arable land by the number of ploughs it needs. The author suggests that these specific measurements may be the result of woodbanks which fix the boundaries, while meadows and arable land have more flexible bounds. More evidence for boundaries is to be found by examining the landscape and parish boundaries. It was in 1955 that W.G.Hoskins made the point that the modern landscape still has traces of past ages imprinted upon it. Past civilisations and modern technology have created the landscape which we see today.

Hoskins believed that while many boundaries were established in Saxon times, some were set by Roman villas, and some, such as double banks or ditches, may be of Celtic origin. Oliver Rackham (1986) says that woods had boundaries marked by woodbanks, constructed to control animal movement, some of which still exist, and may be as much as forty feet in total width. He writes of woods in the south of England and mentions Canfield Hart (Essex) and its neighbouring woods as having boundary ditches, of dry stone walls facing or replacing woodbanks in the Mendips, and of Cheddar Wood as having had a bank to which a wall was added later

Boundaries of woods were substantial in order to keep out deer, a height of about two metres being necessary. Deer browse on young shoots and this ruins all plans for managed woodland. Lyme Park, Cheshire, well known for its deer, has dry stone walls (shown in the photograph, Plate 1:1.) and although paucity of documentation for this area precludes evidence from Saxon times, its earliest date of 1359 indicates a well established estate (L.Cantor, 1983) Oliver Rackham points out at the woodbank is traditionally set with a hedge and a ditch on the outerside to keep out livestock and that the legal boundary of the estate, often on one side of the wood, is marked with pollarded trees as depicted in Figure 1:1.

Some woods and copses recorded in Domesday are of irregular shape .This may be due to the contours of the land, the steep sides of a valley making it unsuitable for farming, as Rackham observes;

Woods tend to been formed, and to have survived, not so much on sites that are good for growing trees as on sites that are bad for anything else.(Rackham, 1976)

Woods survive on flat clay hilltops which were difficult to drain, and on slopes too wet or steep to cultivate.(Rackham, 1986)

Sometimes a boundary ran through a much wider wood, as the eleventh century charter of Compton Greenfield, Gloucestershire,(S. 1362) suggests:

healf þone wudu þærto
half the wood in addition.

The accompanying photograph (Plate 1:2) shows a boundary bank in Hockley Wood, Essex. Woods were valuable assets and were marked by out and looked after accordingly.

In woodland, most parish boundaries are marked by banks; in effect they create two separate woods. But there are several ancient woods divided by well -documented boundaries of which I can find no trace on the ground.(Rackham, 1986)

This observation by Rackham indicates that the ravages of time may have obliterated boundaries, but the documentary evidence implies that woodland was owned by individual estates, and that some woods were divided by boundaries.

The cutting of wood is illustrated in the eleventh century B.L.Cotton MS. Tiberius B.v., fol.6. (see Figure 1:1) and B.L.Cotton MS. Julius A.vi.,fol.5v. where the two interrelated pictures show wood being loaded onto a cart and men cutting trees with axes. One of the trees shows a swelling at the top of the trunk, typical of the shape developed by pollards.

Figure 1:1

Woodcutting

This drawing is based on a calendar depicting the tasks of the year, under the heading 'July'.

Redrawn from P.McGurk, D.N. Dumville, M.R.Godden and Ann Knock, *An Eleventh Century Illustrated Miscellany,* E.E.M.F. 21, 1983.

Plate 1:1

Wall at Lyme Park, Cheshire. OS Ref. SJ963 818.

Although this boundary wall, built to protect the wood from deer, is not of Anglo-Saxon date, it probably replaces one which was built when the park was first established. The dry-stone walls which surround the woods on this estate are dictated by shallow soils and an abundance of surface stone in the vicinity.

Plate 1:2   **Boundary Bank in Hockley Wood, Essex** O.S.Ref. TQ 834 923

This internal bank marks the abuttal of two properties in Hockley Wood . A notice board at the entrance to the wood states that this wood is thought to be part of the remnants of woodland established at the end of the last ice age. The internal boundaries probably date from Anglo-Saxon times.

(5) Crops

Rackham's definition referred specifically to the crop of wood from coppiced woodland, which is the crop harvested today. In the Late Anglo-Saxon Period woodland was also used for pasture and pannage as the Domesday entry for Faversham, Kent, indicates:

> Silva. c. porc'. *et* depastura silva xxxi sol'. et ii den',.
> Woodland,100 pigs; from woodland pasture, 31s.2d;
> ((Philip Morgan,(ed.) 1983)

Rackham makes the point that the pannage indicated in Domesday is a measure of the size of the wood rather than the ability to feed pigs (Rackham, 1976). Constant grazing by pigs would destroy new growth and so destroy the woodland. From this we may infer that *silva porc'* was woodland composed of standard trees, and that dens may have been fenced to prevent damage to adjacent woodland, which would clarify the lawsuit referred to below. The production of acorns, known as mast , is unreliable, and so the value of the woodland varies. A charter dated c.1050 (S.1555) refers to mast as a sporadic crop:

> mæstenrædene þonne mæsten beo
> the right of having mast when there is mast
> (Robertson,1939)

There is also a record of a lawsuit (S1437) of 825 about woodpasture being extended.

The Domesday entry for Pembridge, Herefordshire indicates the unreliability of mast with the entry

> Silva ibi erat ad. CLX. porc' si fructificasset
> There was woodland there for 160 pigs, if it had produced.
> (F & C.Thorn (eds) 1983)

The unreliability of the mast crop is also hinted at in the Domesday entry for Ulverley Warwickshire.

> Silva. iiii. legu' lg'.et dim' leu' lat'. cu' onerat val'. XII sol'
> Wood four leagues long and half a league broad is worth twelve shillings when it bears.
> (John Morris (ed.) 1976)

This makes sense when we read in *The Forester's Handbook* that oak *Quercus spp.* and beech *Fagus sylvaticus* bear abundant seed only in 'mast years', which occur perhaps only once in three or four seasons.It would appear that the value of the woodland varied according to the amount of mast produced. In the majority of references the value of the woodland is included in the value of the estate. In the instances specified in the appendix the value is given for 'when it bears'.The translations variously give 'when exploited','when stocked', but the Latin 'oneratum' is given the meaning 'load, burden' in Charlton T. Lewis & Charles Short's Latin Dictionary.

The use to which woods were put is recorded in Domesday as 'minutæ. silva' (coppice) (F.& C.Thorn (eds.) 1979) and 'pasnag silvæ' (pasture wood) (John Morris (ed.)1976). The charter referring to the lease of Elmstone Wood (see above) includes the clause:

> ða wudu-raeddenne in ðam wudu ðe ða ceorlas brucaþ.

> The right of cutting timber in the wood which the peasants enjoy.
> (A.J.Robertson (ed.)1939)

Domesday also records the industrial use of woodland, such as the supply of cartloads of wood to Droitwich saltworks.(Rackham,1976) Charcoal production is referred to in 969 (S.772) at Apsley Guise 'The old coal-pit where the three boundaries go together.' (A coal-pit in Bedfordshire has to be a pit for making charcoal. (Rackham, 1976)
Timber is the larger wood obtained from mature trees known as standards. The underwood is the coppiced wood which is harvested on a rotational basis and used for wattle, baskets and charcoal.The eleventh century [960-1060 (1025?)] 'Rectitudines Singularum Personarum' refers to the woodward and this further implies that woodlands were managed and treated as a renewable resource.

The woods were also protected by legislation as the ninth century laws of King Alfred, which incorporated seventh century laws of King Ine, show.

> (43)Ðonne mon beam on wudu forbærne, *and* weorðe yppe on þone ðe hit dyde, gielde he fulwite: geselle LX scill.:for ðam ðe fyr bið þeof.

> If any man burns down a tree in a wood and it becomes known who did it, he is to pay full fine; he is to pay sixty shillings for fire is a thief.

> 43.1 Gif mon afelle on wudu wel monega treowa, *and* wyrð eft undierne, forgielde III treowa ælc mid xxx scill.; ne ðearf he hiora ma gelden, wære hiora swa fela swa hiora wære: forþon sio æsc bið melda, nalles ðeof

> 43.1 If anyone fells in the wood quite a number of trees, and it becomes known, he is to pay for three trees at thirty shillings each; he need not pay for more of them however many they were, for the axe is an informer, not a thief. (Laws of King Ine.)

(King Alfred's Laws further state;

> 12 Gif mon oðres wudu bærned oððe heaweð unaliefedne, forgielde ælc great treow mid V scill., *and* siððan æghwylc, sie swa fela swa hiora sie, mid V pæningum *and* xxx scill.to wite.

> 12 If a man burns or fells the wood of another, without permission, he is to pay for each large tree with five shillings, and afterwards for each, no matter how many they are, with five pence; and thirty shillings as a fine

> 13. Gif mon oðerne æt gemænnan weorce affelle ungewealdes, agif mon þam mægum þæt treow, *and* he hit hæbben ær xxx nihta of þam lande, oððe him fo se ðe ðone wudu age.

> 13. If at a common task a man unintentionally kills another [by letting a tree fall on him] the tree is to be given to the kinsmen, and they are to have it from the wood within thirty days, or else he who owns the wood is to have the right to it.
> (Liebermann (ed.) 1903)

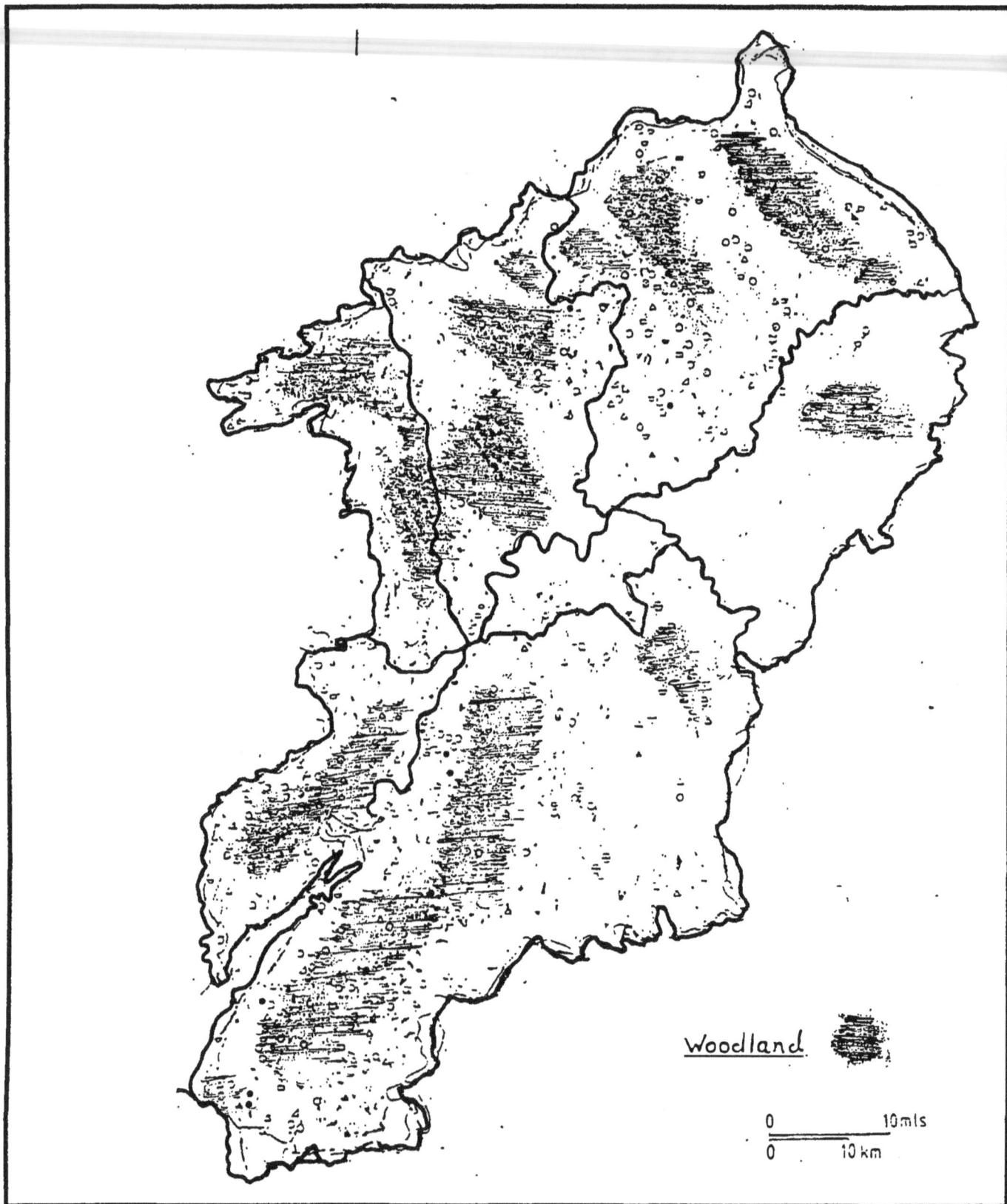

Figure 1:2

The distribution of Woodland in the West Midlands. Adapted from Della Hooke(1981)

It is not easy to burn a living tree, and these provisions probably apply to uncontrolled assarting (clearing wood by burning).

There is some evidence for assarting in place-names. Swithland (Leics), Barnet (Herts and Middx), Brentwood Essex) and Brindlet (Cheshire) all refer to burnt wood-land.(Hoskins 1955)

Hoskins sees England in Saxon times as:

> Still thickly wooded, even in districts that had long been settled. Generally it was thick oak and ash forest.

Place-name elements give evidence for a wooded land-scape, often by indicating a glade or clearing, as in leah which has widespread usage, and a variety of applications, which are discussed by Della Hooke in *Anglo-Saxon Landscapes of the West Midlands: the Charter Evidence*. The meanings, briefly, are 'a wood, woodland', 'a glade', 'a rough clearing in a wood', 'a cultivated or developed glade or woodland clearing', and in later times 'a piece of open land or meadow'. A.H.Smith (ed), (1956) records that the element holt is associated with a wood, thicket, and occurs in Dorset, Norfolk, and Hampshire. The element hurst meaning 'a hillock, a copse' is found as a compound as in Deerhurst, Glos., and Ashurst, Kent. A.D.Mills.gives Shaw (sceaga), 'a small wood, copse, strip of undergrowth or wood' is often found as a simplex, as in Shaw, Greater Manchester, and Wiltshire, and also as part of a compound, as in Shawbury, Shrops. Smith, (1956) makes the point that the Place-Name Elements græf, graf and græfe are difficult to distinguish: græf indicates a grave or digging, the others indicate a copse, thicket or grove.

Della Hook has plotted topographical place-names and charter references to woods of the West midlands and her results are summarised in Figure 1:2. The distribution of woodland is indicated, and swathes of arable land are clear-ly shown. One can see that woodland would not always be near the estate to which it belonged, as a charter of 996 (S. 887) for Benson, Oxfordshire states:

> ðis sind ðas wudes gemære ðe to ðam land gebriad
> These are the bounds of the wood that belong to the land.
> (Kemble (ed.) 1839-1848)

There then follows a perambulation for detached territory some miles away in the Chilterns. (Rackham, 1976) The woodland boundaries, now surviving in places as parish boundaries, and the laws of King Ine and King Alfred indi-cate control of the woodland as a valuable resource.

The demand for domestic firewood, both for heating and cooking, was constant and the industries of iron smelting, (using charcoal), and leather, (using bark and wood ash) and glass making (using wood ash and wood for fuel), made large demands on the woodlands.

Management of the Woodlands

The usual method of management is coppicing, which uses the natural regeneration of trees. (see Plates 1:3 and 1:4.) Broadleaved trees will regenerate after being cut down, and, properly managed, a crop of wood can be gathered at intervals of five to twenty years. This gives a constant sup-ply of wood, of comparable size, although there will be variations in diameter due to varying degrees of light and nutrients. Coppicing has been practised for thousands of years, as the excavation of the Sweet Track in the Somerset Levels shows. Although this is of Neolithic age the con-struction of the hurdles indicates well established coppic-ing.(Rackham, 1977) Some trees are allowed to grow to maturity, and this is known as coppicing with standards. The advantage of this method is that the felling of mature trees causes a minimum of damage to neighbouring trees. The size of the mature tree influenced the planning of large buildings, as internal supports would be needed in very wide structures. As will be shown later, the Anglo-Saxon carpenters were superb craftsmen, and, as well as producing work of a high quality, were supplied with timber which was well seasoned and grown in managed woods. There were, as now, trees which were not in a wood, but in hedges or open grassland.(Rackham,1986) These were accessible to deer and so they were pollarded, cut to a height just out of reach of animals, so that the new shoots could be harvested in the same way as coppiced material.

The conditions under which different trees grow vary. Willow, *salix spp.*, which is coppiced for withies for bas-kets, hurdles, and wattle work, needs wet soil where the water is moving; stagnant conditions will kill the tree. Ash, *Fraxinus excelsior*, grows best on deep loams over lime-stone or chalk. Poorer quality ash will grow on reasonably fertile well-drained soils, but not on podsols, sand or peat. Beech, *Fagus sylvaticus*, and oak,*Quercus spp.*, grow on clays.(Edlin, 1953) This would dictate the establishment of specialised industries, such as basket making in areas of willow, *Salix spp.*, or charcoal burning in areas of oak *Quercus spp.*and beech, *Fagus sylvaticus*.

The occurrence of place-names such as Withington (Cheshire, Lancashire, Herefordshire and Shropshire) (Mills, 1991) and possibly Wythenshawe, (Manchester) and Withy Grove (Manchester) point to areas of coppicing of willow, and possibly local industry. One should not, howev-er place absolute reliance on the modern form of a place-name, as the Withington in Gloucestershire was Wudiandun in 737 and Widindune in 1086 (DB.) (Mills, 1991) The evi-dence of boundaries to woods, protecting the crop by hedges, ditches, banks and walls, internal boundaries indi-cating the demarcation of properties and legislation to pre-vent wholesale destruction, all indicate careful woodland management by the woodward.

Pannage was a major use of the woodlands in the Late Anglo-Saxon Period, as the many Domesday entries referring to 'silva porc' show. The evidence which Della Hooke has presented in *Anglo-Saxon Landscapes of the West Midlands* (1981) and the charter for Benson (Oxfordshire) (S. 887) quoted above indicate the need for transhumance while evidence given by K.P.Witney in *The Jutish Forest* shows how dens were held in the Kentish Weald, which were used for pannage during the autumn:

> The specialisation in land use exacted its own penalty. In particular it necessitated a great annual transmigration of stock to their fattening grounds. Moreover the pannage season lasted for some seven weeks

Woods were also the habitat of wild life and so were used for hunting, as the laws of King Cnut allowed;

> (80) And ic wylle, þæt ælc man si his huntnoðes wyrðe on wudu *and* on felda on his agenan.
> I will that every man be worthy of his hunting in wood and field on his own estate.

> (80, 1) *and* forga ælc man minne huntroð, loce hwar ic hit gefriðod wylle habban, be fullan wite.
> And let every man abstain from my hunting: look wherever, I will that it should be freed under full penalty
> (Liebermann, (ed.)1903)

The hunter in Ælfric's *Colloquy* states:

> Ic gefeo heortas *and* baras *and* rann *and* rægan *and* hwilon haran.

I catch harts and boars and roe deer and does and sometimes hares. (Garmonsway, 1939)

The Domesday reference for Longdendale, Derbyshire, refers to the area as fit only for hunting.

> Wasta. e tota Langedendele Silva. e ibi n pastil. apta venationi. Tot'. VIII. le'u lg. and iiii. leu' lat' T.R.E. XL sol'
> All Longdendale is waste; woodland, unpastured, fit for hunting. The whole 8 leagues long and 4 leagues wide. [value] before 1066, 40s. (Morgan ,1978)

The uses of the woods, while not always meeting Rackham's definition of producing a crop of coppice, are stated in Domesday as being 'silva pastilis,' 'silva minutæ' or 'silva pasnag'. In Yorkshire the entry for Borrowby and Roxby is

> 'silva ñ past'
> woodland not pasturable
> (Faull & Stinton)

The information which we have about the protection of woodlands by boundary banks, legislation and woodwards indicate the management of a valuable resource. The industries dependent on the woodlands show just how important the woodlands were to the Anglo-Saxons.

Plate 1:3   A newly planted coppice (above)
Plate 1:4   Regrowth on a stump (below)

Part Two

The Coppice Industries

The use of wattle in the Late Anglo-Saxon Period

Before discussing the use of wattle it will be useful to distinguish between wattle and basketry. Wattle is the weaving of osiers from coppiced willow or hazel and may be used as part of a structure. Baskets are made from the same material and are used as containers. It has been shown that wattle was used in Neolithic times,(Rackham, 1977) its earliest known use in England was as reused hurdles in track-ways in the Somerset levels and the associated coppicing provided a constant supply of material. The advantages of wattle are the use it makes of young growth, which means greater production from a small area, ease of cutting as compared with the felling of standards and the relative strength of the finished product.

Wattle hurdles are made by pushing the uprights (known as sails) into the ground or a horizontal frame and weaving the horizontal members (known as rods) into them. Details of a hurdle are shown in the photograph, plate 2:1. In the case of hurdles they are then removed from the ground or frame to be used where required. In some cases a loop is made at the end of the topmost rod and the opposing sail is made longer so that hurdles can be locked together when structures such as sheep pens are needed. The Old English word for this was lochyr dle.(Bosworth & Toller,sup,1921) For more permanent structures, such as walls, wattle is made *in situ*.

Wattle was used in the late Anglo-Saxon Period as in-fill in timber framed buildings, just as King Alfred prescribed 'weave many elegant walls' and as internal screens and walls. Hurdles were used, among other things, in fish weirs. (F.M. Losco & O. R. Bradley, 1988) There is abundant evidence from Anglo-Saxon England for the use of hurdles, particularly for fishing. The Old English words *cyter* 'a basket weir', *haccwer* 'hack weir' and *wera* 'a fixed structure for catching fish' confirm the use of weirs made of wattle. Domesday Book records many fisheries, for example, two are mentioned at Duncton, Sussex.

> ii piscariæ de. ccc. lx, anguill.
> Two fisheries at *360* eels
> (Morris, 1976

Another entry, for Kingston, Surrey, records,

> ii piscari de. x. sol. *et* tercia piscaria valde bona. sed sine censu.
> Two fisheries at 10s; a third fishery excellent, but without dues.
> (Morris, 1975)

Place-names are discussed by V.E.Watts in 'Medieval Fisheries in the Wear,Tyne and Tweed: the place-name evidence'.The element 'yair' is defined as 'an enclosure commonly of a semi circular form built of stones or constructed of stakes or wattle work, stretching into a tideway, for the purpose of detaining fish when the tide ebbs.' (V.P. Watts, 1983) The O.E. *hæc*, 'a hatch, a grating, a sluice gate', occurs in *Hachesiare* (1195) meaning a 'yair with hatches or called Hatches', so that a barrier or fence of wattle set across the current is implied. This created an eddy in which fish could be caught in a boat.

The site of a fish-weir was described by Losco-Bradley and Salisbury in 1988. The River Trent meanders at Colwick and the extraction of gravel from the old river bed exposed an eleventh century Anglo-Saxon fish-weir. The Colwick discovery was made in alluvium in which only deeply buried objects, or posts driven in to the mudstone, remain *in situ*.The excavation revealed hurdles in a horizontal position and posts, some horizontal, some broken off in the river bed, which gave enough evidence to show that the minimum length of the structure was some thirty five metres. The wattle was held in place by a double row of oak posts, of 11·5 to 14cm. in diameter.The posts holding the wattle at Colwick seem to have been buttressed by more posts on the downstream side, held by boulders at the base. Examination of the hurdles showed them to be made of material from different sources. One was made from coppiced hazel from a stand with no rotation of cuttings, a system known as 'drawing'. The rods varied in thickness from 10 to 20 cm. and had an age range of about six years. A second hurdle was made of willow which had been both coppiced and pollarded. The pollarding is indicated by narrow growth rings caused by 'topping' when growth continues by a side growth, producing scars. An incomplete last ring indicates summer cutting, so necessary for hurdles being ready for catching eels in the autumn. Topping and summer cutting together suggest the harvesting of leaves and shoots for cattle fodder, after which the stems were allowed to grow for a second harvest when they were cut for hurdle making.

The construction of the hurdles was on 'sails' or uprights set twelve to thirteen centimetres apart, each panel being about fifty centimetres high. The age range of the sails was twenty seven years, again indicating the 'drawing' system of coppicing. The plan of the weir was that of a funnel pointing downstream, which would draw fish and eels into baskets at the apex.

There is a caveat to the effect however that although some stake alignments may be fish weirs, the majority of finds may be river bank revetments, (Losco-Bradley & Salisbury, 1988) An example of this use was found in the excavations carried out by the Extra-mural Department of Manchester University, under the direction of Dr. David Hill, at the site of Quentovic, France, in 1990. At Quentovic the wattle was preserved in a bank of water-borne sand because the water table was high enough to preserve the wood in anaerobic conditions. The accompanying photograph, Plate 2:2, shows the remarkable state of preservation and the depth to which it was buried.

The use of wattle in buildings has been found in Dublin and York, and so the available archaeological evidence is from the two cities of the Viking kingdom of Dublin-York, on the periphery of Anglo-Saxon England. The evidence is an indication of what was probably widespread practice at that time. The environment in both cases is anaerobic and so organic remains are preserved.

There are four sites in Dublin, and they are thought to represent the second Scandinavian settlement of the tenth century. (The first settlement of the ninth century is thought to have been higher up the River Liffey.) (Hilary Murray, 1983). Rescue excavations were carried out in 1962-63 and 1967-76 by A.B.O. Ríordáin in High Street, Winetavern Street and Christchurch Place, Dublin.

At this time there was little experience of excavation in waterlogged sites, and so not all details such as tool marks were recorded. Among the problems encountered were posts which had been forced up through overlying layers as the ground compacted. The subsequent displacement made interpretation of evidence difficult. The excavations revealed post and wattle walls, which were of three basic types. Single post and wattle, double post and wattle and wattle with planks. A fourth method of construction was of staves set in a sill beam. It is suggested that double wattle walls were insulated with material such as dried cow dung, mud or turf. (Murray, 1983)

Coins and dendrochronology gave dates of twelfth to fourteenth centuries. (Murray, 1983) Few earlier structures had timbers suitable for this type of dating, which relies on the growth rings of oak. In most earlier cases ash was predominant, supplemented by reused timber in a scrappy condition. In these cases the context could only be given as pre-twelfth century. The structure of the wattle was of sails made from round posts, generally retaining their bark. The ends were pointed, probably with an adze, and some were fire hardened. These had been hammered into position and the horizontal rods were then woven in. The size of the horizontal rods varied according to the size of the posts, the whole structure indicating coppicing. (Murray, 1983) A final note on the use of wattle in Ireland is found in the Gaelic name for Dublin *Ath Cliath* meaning 'the ford of the hurdles'. (P. F. Wallace & R. O. Floinn. 1988).

The excavations at York were made at a later date (1972) than those at Dublin and so more details were recorded. The conclusions were however much the same: wattlework buildings from the first half of the tenth century were repaired or reconstructed several times, and there was evidence of repeated gutting by fire. The buildings were from 3·5 to 4 metres wide and of unknown length, as the area of the excavation did not reveal the complete plan. They were however in excess of 8·2 metres in length. (Hall, 1982) Wattle was used in internal furnishings as a revetment for an earth-filled bench. Although daub was found in and around the buildings, none was found actually on the wattle. It has been suggested that the wattle tradition was abandoned during the decade 950-960 when horizontal timbers were introduced. Planks some six cms thick were set on edge and stakes were set at 25 cm intervals to form the rear wall. The structures appear to have been sunk into the earth as cellars, some 1·5 m below the contemporary surface. The foundation beams had a lip to prevent the uprights from moving inwards. In one case there was evidence of a cavity wall, charring on the inside surfaces indicating free circulation of air in the cavity. Drains were lined with wood or with wattle to go round corners.

Wattle was also used as shuttering in the construction of concrete or plaster details within stone churches, and the impressions remain in window openings at Hadstock, Essex, (now hidden because of boarding up) and at Hales, Norfolk. (H.M.Taylor & J.Taylor, 1965) The sails of the wattle were inserted into holes drilled with a spoon bit in the window frame and the wattle structure held the rubble in place until the mortar had set. Modern spoon bits are shown in the photograph, Plate 2: 3, and were used up to the advent of power tools. The wattle must have been woven onto the window frame before the latter was placed in position, making the whole assembly an unwieldy piece. It would only be practical to position it accurately from scaffolding. Wattle is a very versatile material which was widely used during the Late Anglo-Saxon period.

Basketwork in Anglo-Saxon Times

The archaeological record does not contain surviving basketry as opposed to wattle, yet its use is undoubted and substantiated by the Old English words *spyrte* (Garmonsway, 1939) a wickerbasket or creel, *sæd leap* (Bosworth & Toller, 1838-1921) a seedbasket, and possibly as an element in *cyter*, a basketweir.

Baskets are illustrated in the Utrecht Psalter (Utrecht University Library, MS 32). This manuscript is believed to have been made at the monastery of Hautvilliers near Rheims in the ninth century. The Canterbury Psalter is based on the Utrecht Psalter, as the Utrecht Psalter was at Canterbury in the centuries immediately following its execution. (P.D'Ancona & E.Aeschliman, 1969). Although Carver (1986) warns about the possibility of the influence of classical tradition, it is unlikely that an artist would draw an object which he did not recognise. In this instance it is the object and not the design which is important. The common usage and necessity for containers in every day life is well known to everyone who picks up a polythene bag, and mundane articles are so taken for granted that they are not written about. We can however infer specialised uses, such as the eel trap shown in the photograph and from the many references to eels in Domesday Book.

Although eels are also referred to in Ælfric's *Colloquy* (G. N. Garmonsway, 1939) and the place name 'Ely' means 'the place where eels are caught', (Mills, 1991) the weight of evidence comes from Domesday Book, where eels are often referred to in relation to mills, as the entry for Alveston, Warwickshire records:

> Ibi. iii molini de . xl, solid. *et* xii. stich anguill. *et* mille.
> Three mills at 40s. and 1012 sticks of eels. (Morris, 1976)

Plate 2:4 shows an eel trap, and an illustration in the fourteenth century Luttrell Psalter, folio 181, shows eel traps in a mill stream, as they were probably used in the Late Anglo-Saxon Period. The use of baskets is referred to in Ælfric's *Colloquy*:

> Ic æstige min scip *and* þyrpe max mine on ea, *and* ancgil *vel* æs ic þyrpe, *and* spyrtan *and* swa hwat swa hig gehæfted ic genime.

> I go aboard my ship and throw my net in the river and throw my baited hook and basket and so I keep what I catch. (Garmonsway, 1939)

Plate 2:1 Hazel hurdle, showing twisted ends, necessary to stop the hazel breaking, and the base in which it is held during construction.
Bill Hogarth's coppice, Map ref. SD 349 861

Plate 2:2   Wattle revetment, identified as part of the water front on the River Canche at Quentovic. This shows the size and strength which it is possible to attain with this versatile material. (Photograph by courtsey of Mrs. Margaret Worthington) June 1990.

Figure 2:1

The use of wattle is illustrated in B.L.Cotton, Tiberius B.V. (Volume 1) and B.L.Cotton,Julius A. VI (f.6v.) where it is shown as hurdles forming the sides of a cart. Forks are also depicted, which are probably made of wood, similar to that in plate 2:6.

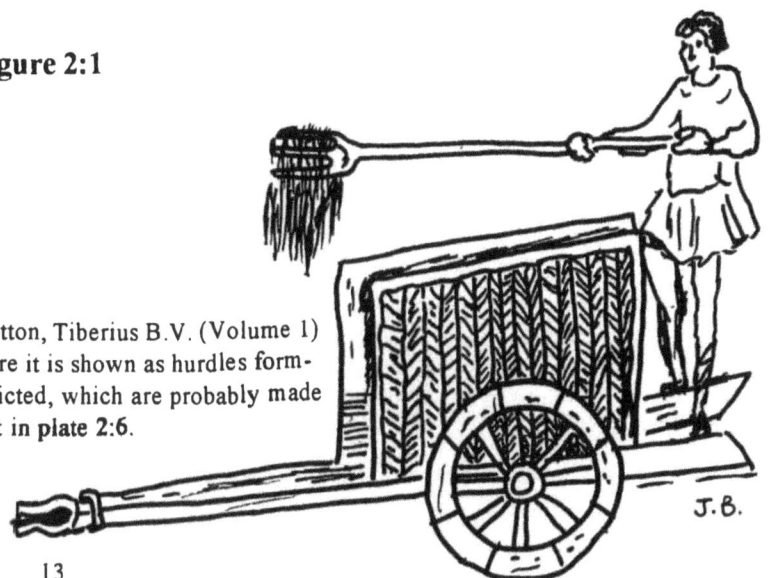

13

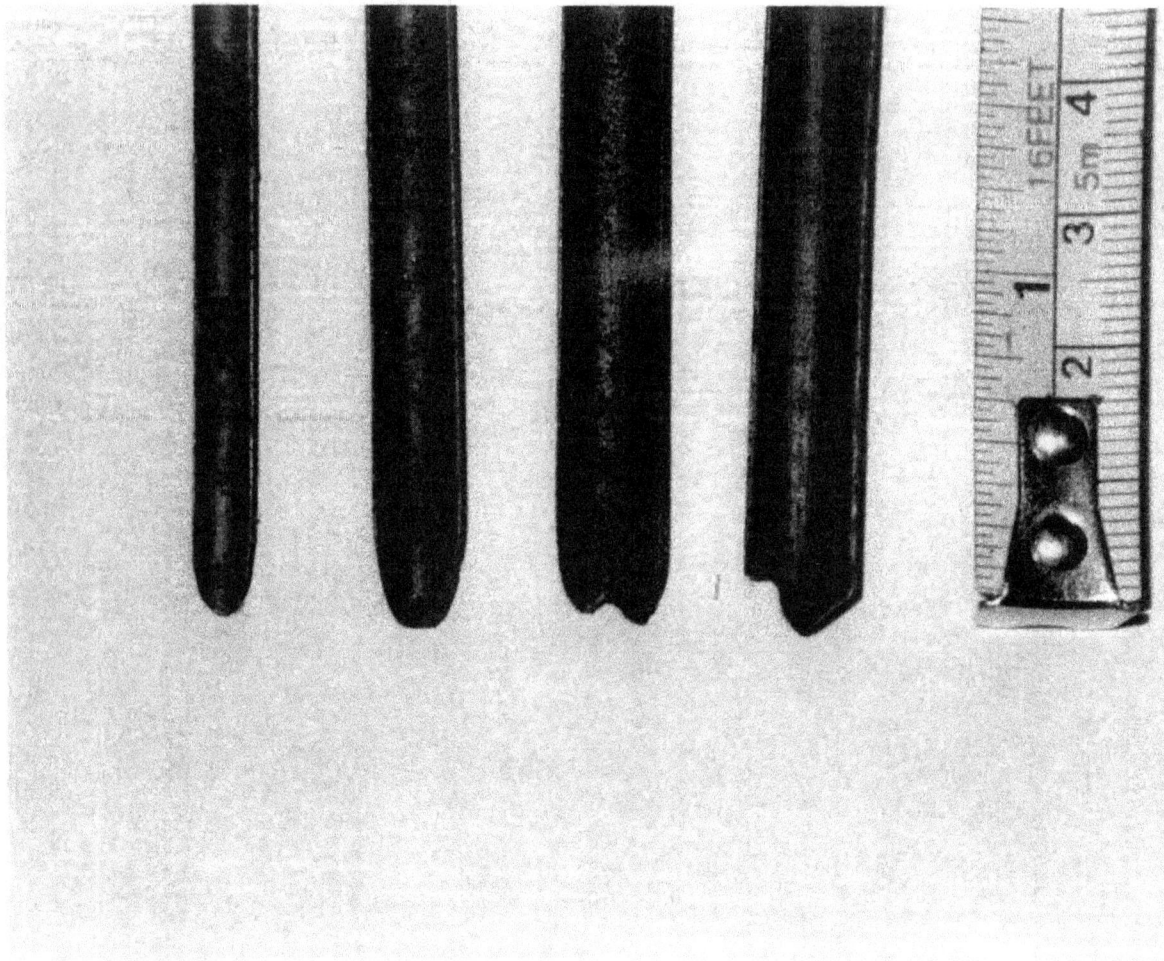

Plate 2:3

Spoon bits, still in use within the last 50 years, were used to drill holes in window frames to hold wattle formers..(the hole has a recognisable shape at the bottom.)

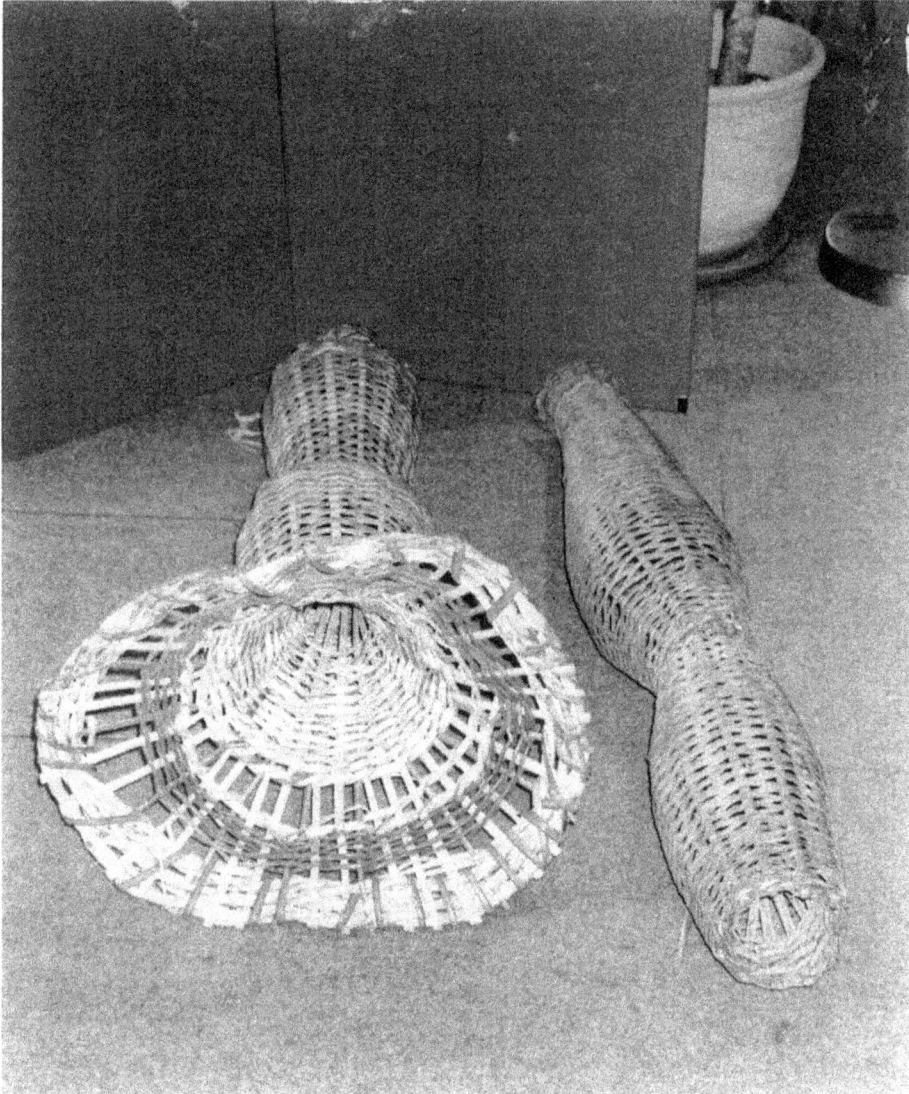

Plate 2:4
  Eel traps at the Archaeological resource Centre,
York. The left hand trap is for salmon, that on
the right for eels.

Basket weirs are also referred to in charters, as in the 'Survey of the Manor of Tidenham,' Gloucestershire, (S. 1555)

> On Sæverne xxx. cytweras.
>
> 30 basketweirs on the Severn
>
> xiiii cytweras on Sæverne. *and* II hæcweras on Wæge.
>
> 14 basket weirs on the Severn and 2 hackle weirs on the Wye
>
> To Cynestune on Severne xxi. cytwera. *and* on Wæge' XII
>
> At Kingston there 21 Basket weirs on the Severn and 12 on the Wye. (Robertson, (ed.) 1939)

Baskets were used as measures in Carolingian Europe, as is indicated in the document known as 'Capitulare de Villis vel Curtis Imperialibus'.

> (9) It is our wish that each steward shall keep in his district measures for *Modii* and *Sectaria*, and vessels containing eight *sextaria*, and also baskets of the same capacity as we have in our palace. (H.R.Loyn & Percival, 1975)

The 'Brevium Exempla' is a model of how an inventory of an estate should be recorded, and baskets are referred to as containing spelt (a type of wheat).

> Produce: Nine baskets of old spelt from the previous year, which will yield 450 measures of flour; In the present year there were 110 baskets of spelt: (Loyn & Percival).

Although this evidence is from the continent and before the Late Anglo-Saxon period, it is close enough in time and space to indicate the widespread use of basket work.

The use of baskets in salt production at Droitwich is suspected by Della Hooke (1981), but once more the evidence is inferential. In her article entitled 'The Droitwich Salt Industry' she refers to the Roman industry and says,

> The final stages of draining the impurities from the salt were later carried out in basketware containers and this method may have been employed in the Anglo-Saxon period.

Figure 2:2

(Eleventh century Canterbury Psalter f.233 b)

Figure 2:3

(Canterbury Psalter f.41b) .

Plate 2: 5

Wooden Fork

Wooden forks such as this at 'La Veille Maison', Aignon, France, are illustrated in Cotton, Tiberius B.V. (volume 1) (f. 5v) and Cotton, Julius A VI, (f6v). They were probably grown as part of the coppice crop, shaped by judicious pruning, and would be in constant demand.

The coppice industries included the manufacture of many small items such as forks, two modern examples are shown here. Today's coppice products include walking sticks and shepherds crooks. We must assume that these and other useful products were made as demand arose.

17

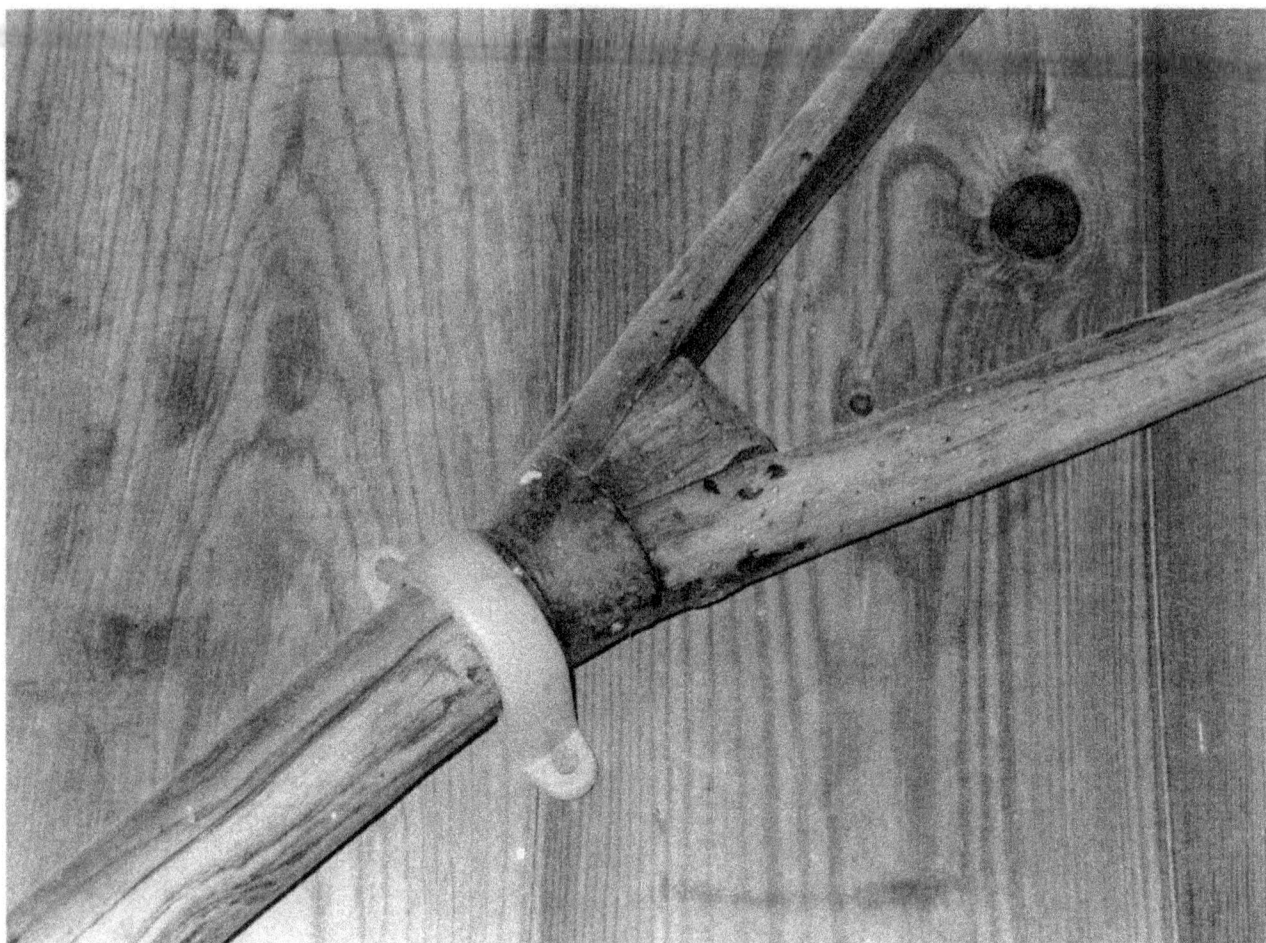

Plate 2:6

Detail of a fork made by splitting a pole. In the Late Anglo-Saxon Period the binding was probably withy or string made from bast. There does not appear to be string made in this way from within this period but it was used in scaffolding rope shortly afterwards. For short lengths a handful of the fibre would be rolled on the thigh in a similar manner to which the Africans do today. It is a short step to make this into a rake by attaching a suitable end.

Figure 2:4

Besom Making

Plate 2:7

Besoms

Besoms are still made today, and are one of the main uses for birch twigs.

Ðanne cwed he, Ic gecherre on min hus þanen ic uteode; and cumende, he gemet hyt emtig, geclænsed mid besum, and gefatewed'

'þæne he cymd he hit gemet æmtig mid besmum afeofmod

Both meaning 'I came to my house and found it to be empty and swept clean'

These two quotations from the Anglo-Saxon gospels refer to besoms as brushes

(Hardwick (ed.), from work by Kemble, 1858) (Matthew 12,14)

(W.W.Skeat, 1874) (Luke 11,25)

Besom Making

Besoms (O.E. besema) are referred to in the late ninth century Old English translation of *The Orosius* , (J.Bately (ed.), 1980), a fifth century Latin work which was translated into Old English and so we know that they were used during the late ninth century.

The method of making besoms is not likely to have changed much since the Late Anglo-Saxon period.

The crowns of birch are cut at about seven years growth as part of the coppicing programme, although other plants may have been used in the past such as broom and heather. The crowns are stored over winter, and are then ready for making into besoms. The most difficult part of the job is holding the bundle which is to form the bosom head. The line drawing (Figure 2:3) shows the use of a 'horse' which holds the

20

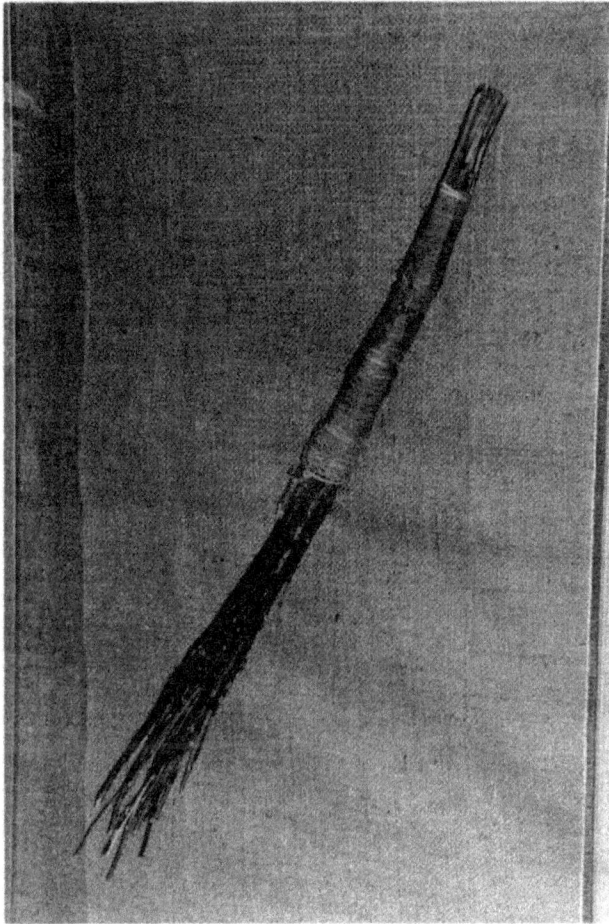

Plate 2:8

Birch

The second meaning of besema is to be found in the Osorius, where it is a means of punishment.

eallum þam folce mid besman swingan
all the folk were beaten with besoms (J.Bately,1980)

The birch was used as a punishment until 1948. As the photograph shows, it is very similar to the other besom.
(Photograph courtesy of Manchester Police Museum)

binding material, generally a strip of hazel or stripped bramble. The handle is pushed into the head, which tightens the cuttings, and it is then 'knocked down' by holding the head and banging the handle on the floor.(J.Arnold, 1970)

A second method involves the use of an 'engine' (see Plate 2:10) which consists of two bars with curved ends which encircle the head of the bosom and are held tight with the knee. The 'horse' would be the more economical to construct in Anglo-Saxon times, as this is made entirely of wood, although the second method would be technically possible.

Plate 2:9

An alternative way of holding a besom head while it is being bound. The two methods, using a
'horse' or an 'engine' are used contemporarily in different parts of the country. They are simple
in concept and either (or both) could have been used in the late Anglo-Saxon period.

Part Three
Standard Trees

Having considered the use of coppiced trees and their products we can now consider the use of the standard trees which were allowed to grow on to provide timber. The principal use was for building construction. (Rackham, 1986) It is likely that house construction was a major use for standard trees in the late Anglo-Saxon period, and the richer the owner the more use would be made of the trees, a poorer house having more wattle. Timber was also used in bridge construction, (one of the obligations of the *trimoda' necesitas)* (F.M.Stenton, 1943) and one structure was of ten spans across the Medway. (Robertson, 1939) (S. 1555).

Timber would be cut when it reached the smallest size required for the structure, thus saving time and energy in removing excess wood. Some of the timber would be required for bowls, cups and other containers. The pole lathe turner bridged the gap between the coppice craftsmen and the user of timber, turning coppiced poles for items such as candlesticks and chairs, and timber for hollow ware. Some containers were too big to be made from one piece of wood, and so were made of staves. Tubs and barrels were the domain of the coopers, who made their containers of oak *Quercus spp.*.

Evidence for the use of the Pole Lathe

Although there are no pole lathes in the archaeological record there are objects which have obviously been turned on a lathe, and Carole Morris has shown that a reciprocating lathe is the only way in which some of the marks on bowls and cups could have been made. [A reciprocating lathe moves with a backwards and forwards motion, as opposed to the more familiar rotary lathe which is driven by a motor.] Chair legs and candlesticks, lectern stands, bed heads and thrones are depicted in the Utrecht Psalter and the Canterbury Psalter of Eadwine. These could be drawings influenced by classical tradition as Carver cautions. (see above) The simpler items are made on a reciprocating lathe, and are known as 'spindle turned'. The more difficult items are 'face turned', and are the bowls and cups which were highly thought of; enough to warrant being repaired with metal plates and rivets, and to receive bronze or silver mounts and rims. (Carole A. Morris, 1982). The pole lathe is still used today by 'bodgers', who work in or near woodlands. This is a craft which had almost died out, but renewed interest has revived the craft, with both professional and amateur exponents. In his research the author has spoken with several bodgers, all of whom speak highly of one man, the late George Lailey, who could turn bowls in a nest, that is one inside another, as shown in Figure 3: 2. This is a skill not easily achieved by modern bodgers, but represents the attainment possible by men whose entire life was spent in the craft, and, by inference, a possible skill of the Anglo-Saxon craftsmen.

The principle of the lathe, illustrated in Plate 3:1, is that a long springy pole is fastened at one end and the other end is free to move up and down. A cord passes from the free end of the pole to the lathe, where it passes round a rod to which the work is fastened. The cord is attached to a treadle which, when operated, revolves the rod, and the pole, in springing back to its original position, turns the rod in the opposite direction. (Mike Abbot, 1989) The wood to be worked on is firmly supported at each end, but is free to revolve. A range of chisels and cutting hooks are used to carve the revolving wood, and these are recognisable by the length of the handle, which may be as long as sixty centimetres, and the wear and grinding of the blade, which leaves the cutting edge at an angle unsuitable for mortise cutting. The tools need a rest to give the operator control.

Illustrations of bowls and goblets which may be made of wood bear witness to the more advanced technique required for hollowed dishes. (M.R. James, 1955) Here the archaeological record substantiates the production of these items, for a turned goblet of yew *Taxus sp.* wood and bowls of ash *Fraxinus spp.* wood have been found at York. In spite of the reservations expressed by Carole Morris about yew *Taxus spp.* affecting the taste of liquids, another cup made of yew was also found at Winchester. (Morris, 1982)

Some woods are more suitable for turning than others, oak *Quercus spp.* being very poor, breaking easily when being worked on. Oak *Quercus spp.* artifacts indicate a high degree of skill achieved by the Anglo-Saxon turners. Good woods for turning and which have occurred in the excavations at York include field maple *Acer campestre,* alder *Alnus spp.,* ash *Fraxinus spp.,* yew *Taxus spp.* and boxwood, *Buxus spp.* A small core of Scots Pine, *Pinus sylvestris* has also been found at York.

The place-name 'Coppergate' is now thought to derive from the ON. kuppr, meaning 'cup', and so the 'Cupmakers Street'. (Morris, 1982). A building containing 'cores' from turning as well as waste and discarded bases, which are cut off the finished product, has also been excavated at York. These bases are readily identified by the marks left by the metal which holds the work in place. Some tool blades and a wooden item which Carole Morris believes to be the 'rest' from a pole lathe have also been discovered.

The wood was prepared by taking a section of tree, which was usually quartered, although sometimes halved, depending on the size of the tree. The base of the vessel was nearly always towards the outside or sapwood of the tree, and the roughout was prepared in this direction. This reduced the likelihood of warpage after turning, which would have split the bowl. The wood was worked 'green', that is soon after the tree was felled, because the wood was easier to cut and shape. The outside of the vessel was left in an almost finished condition, but not much, if any, of the inner was removed at this stage. The roughout was then left over winter to season before being finished. (Morris, 1982). The selection of the direction of the grain and position of the

Plate 3:1

     Pole lathe

Figure 3:1

Examples of 'spindle' turning on lectern and throne.
Detail copied from Canterbury Psalter f.135 (eleventh century)

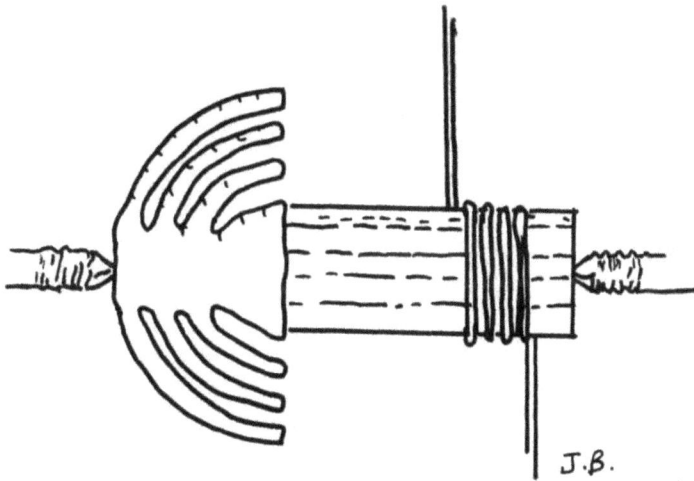

Figure 3:2   Turning a 'nest' of bowls

sapwood indicates an understanding of timber which enabled the Anglo-Saxons to use the material to the best advantage. No roughouts of the late Anglo-Saxon period have yet been found, and so evidence must be sought from other sites. Roughouts have been found in contemporary Irish crannogs and in undated contexts in Irish bogs. Some of these show evidence that part of the inside of the vessel had been removed with a chisel or hand adze.

In summary, we have a picture of craftsmen working in urban workshops, and possibly in woods in the manner of the bodgers. The work was valued, and probably provided much of the table-ware depicted in the manuscript illustrations, such as goblets and cups, and the turned legs of chairs and lecterns. Coppiced poles would be used for spindle work, and standard trees would be used for bowls. The pole lathe worker formed a bridge between the men who cut and used the underwood and the carpenters who used the mature standards.

Coopering

Barrel making is a craft which has survived with very little change, although there are very few coopers in these days of metal drums. The archaeological record has a barrel lined well at York (Hall, 1982) and barrels are shown in the illustrations of Psalm Four in the Utrecht (E.I.De Wald, (undated)) and Canterbury Psalters (M.R.James, 1935). A barrel is also depicted in the Bayeux Tapestry. (L.Musset, 1989) The survival of the craft depended so much on woodworking skills, and has not had any mechanisation, that it is likely that the craft is still as it was in the Late Anglo-Saxon period. Barrels are designed to hold liquids, and at the same time be able to take the stresses of rough handling. The waist enables the barrel to be rolled, and the hoops keep the staves in place. Although we normally think of barrels as holding liquids (wet coopering) they are also used for dry goods, and 'white coopering', which does not necessarily use oak, was used for dairy and domestic items such as pails. (Jack Hill, 1976).

Although oak *Quercus spp.* was the usual wood, chestnut *Castanea sativa* was occasionally used and the successful cooper had to fully understand the structure of the wood. The heartwood was used, all the lighter coloured sapwood being removed. The staves were cut with the growth rings going from front to back, and the medullary rays, which are impervious, went across the width. The staves were stacked in the open to season before being shaped. The slight curve on the outside and inside was shaped to fit when bent by removing wood from top and bottom on the sides. The sides were then bevelled. All this was done by the judgement of the cooper, there being no measurements or templates. The barrel was assembled by placing the staves in a circle and a hoop round the outside. The inside was wetted and the barrel placed over a small fire on a trivet. The heat and steam softened the wood sufficiently to enable hoops of smaller diameter to be forced into position. This process was repeated at the other end, and the bung hole was cut before the 'heads' were put on each end. This meant cutting a groove for the heads to slot into, removing a hoop to allow the staves to open a little, and caulking the groove with a rush before putting in the head and tightening up the staves again. The hoops were made of hazel. (Hill, 1976).

Buildings

The form of Anglo-Saxon buildings has been the subject of debate and conjecture for a long time, and in 1896 Addy postulated a wooden building with 'a ground plan—of a basilica with nave and aisles' (see Figure 3:4). This was based on 1882 accounts by R.Henning and August Meitzen of a form existing in Friesia and Saxony at that time. (S. O. Oddy, 1898). In 1913 E. T. Leeds said,

> They left no monuments in stone like the Romans, simply because they did not understand the workings of stone; they came, for the most part, of a race which has always excelled in woodwork. All their houses, certainly those of any size, were constructed entirely of wood, like the hall at Heorot, so vividly described in Beowulf. (E. T.Leeds, 1913).

C.A.Ralegh Radford commented in 1957 that the settlements on the Continent in the second to fifth centuries had aisled longhouses with two rows of posts carrying the roof, with outer walls of wattle work. This was at Tofting, on the estuary of the Eider, Holstein, an old Saxon area. In sixth century England wooden halls were built as the excavations at Cowdery's Down, Basingstoke, Hampshire have shown. The size of the halls at Cowdery's Down ranged from 5:4 metres x 5·4 metres to 22:1 metres x 8·8 metres, The method of construction was sophisticated, and indicates the high standard of craftsmanship which was attained before the Late Anglo-Saxon Period, (M.Millet & S.James, 1978-9). Other smaller buildings have been found, characterised by a sunken floor. These are known as 'grubenhausen' and for a time were thought to be the homes of the Anglo-Saxons of the Settlement Period. Two were found at Cowdery's Down, and it is now thought that they were used as workshops, stores or even accommodation for slaves.

Wooden Churches and The Church of St. Andrew at Greensted

The tradition of wooden churches is documented by Bede, who refers to wooden churches at Lindisfarne and York, (L. Sherley-Price, 1955) and William of Malmesbury suggests that King Alfred's church at Athelney monastery was also of wood. (N.E.S.A.Hamilton (ed.), 1870) A wooden church at Bury St.Edmunds is referred to in the *Memorials of St.Edmund's Abbey* (Thomas Arnold (ed.), 1870) and a charter (S. 909) of King Ethelred, dated 1004, renewing the privileges of the monastery of St Frideswide, Oxford, refers to the building being set on fire during the massacre of the Danes on St.Brice's Day, 1002. (F.Barlow (ed.& tr.), 1962).

The archaeological evidence includes postholes and slots at Potterne, (N. Davey, 1964) Thetford (B. R. Davison and R. Mackay, 1971) and Elmham (S. E. Rigold, 1962-63).

Plate 3:2

The Church of St. Andrew, Greensted
The fillet of wood inserted between the half trees to give stability and prevent draughts.

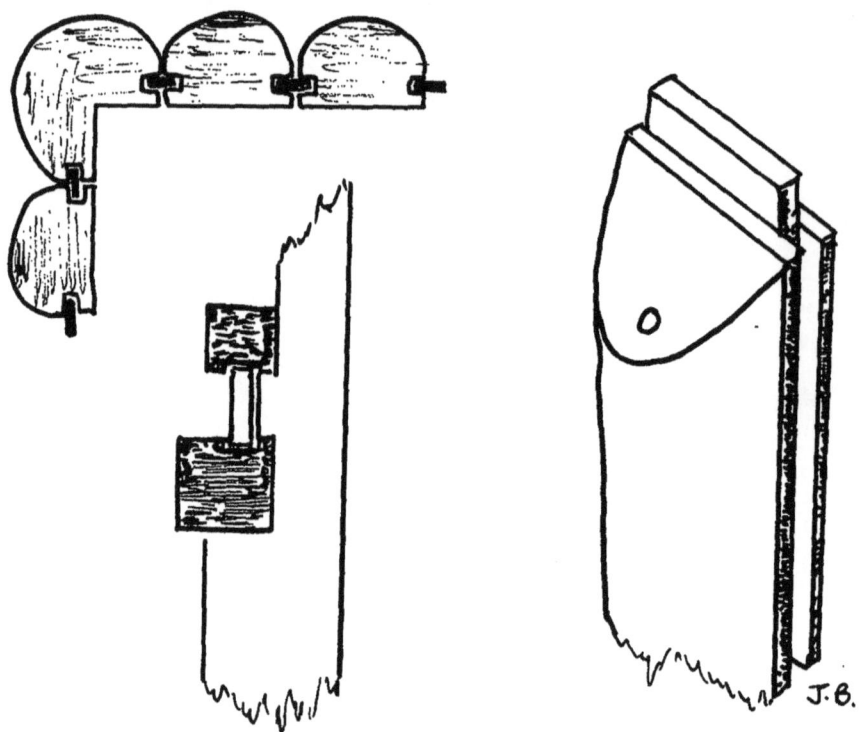

Figure 3:3

Details of the wall joint at Greensted Church

The church of St. Andrew at Greensted, Essex, has surviving wooden walls which have been dated by dendrochronology to 845 A.D. This is recorded in the *Guide to Greensted Church* (Anonymous, undated) but the article in *The Antiquaries Journal* (Hakon Christie, Olaf Olsen and H.M.Taylor, 1979) make the observation that the method by which the dating was obtained was not very efficient and it should be repeated.(An eleventh century date has since been given, author, 1998). Prior to this dating the church was thought to have been built as a resting place for the remains of St, Edmund when his relics were translated from London to Bury St.Edmunds in 1013, (Taylor & Taylor, 1965).

The structure is interesting for two constructional methods. The timber walls were mortised into oak sills, and each stave was married to its neighbour with a narrow strip of wood some two or three centimetres thick and between ten and sixteen centimetres wide which runs the whole length of the stave, (Christie, Olsen and Taylor, 1979). This added stability to the wall and made it draught proof. It can be seen in places where the staves have shrunk a little to leave a slight gap, as the photograph, Plate 3:2, shows. The edges of the staves were planed to give a good edge to edge fit, and the fillets would not have been visible. The fillets were revealed in 1848 when renovation was carried out. The oak sill was removed and a brick sleeper wall put in to carry the staves. The work was reported in *The Builder* in 1849 and it was stated that the walls were of oak about six feet high, including sill and roof plate, and were formed of half trees. These averaged twelve inches by six inches. The oak sill was mortised to receive a tenon of 1½ inches. The sill on the southern side lay on the earth, and that on the north had flints under it in two places. The north west corner of St.Andrew's is a log stave with an inner quadrant removed. Whether this is original or a later refinement is difficult to ascertain, but the task was not an easy undertaking. Although doubt has been expressed about the age of the sill (Christie, Olsen & Taylor, 1979) (which could have been inserted at a later date), the undoubted age of the walls and their method of construction, shown in Figure 3:6, indicates the skill of the Anglo-Saxon craftsmen. The building has been compared with other wooden churches in Norway and Denmark, and although the Scandinavian examples are younger the basic concept is common to all. (Christie, Olsen & Taylor, 1979).

The use of split standard trees for the walls shows the demands of major buildings. The surviving portions of thirty trees (sixty halves) indicate the great amount of timber required for the walls. Roofing timbers and tie beams would account for more standard trees, as well as coppiced poles.

The church would have been fairly draught free, and the natural insulation of wood would have given a warmer interior than stone churches.The article in *Antiquaries Journal* concludes with the following summary:

> The surviving wooden fabric of the nave at Greensted has no direct analogue in standing build-

ings either in England or on the continent; but excavations have shown evidence for the former existence in both places of buildings with walls of upright logs, some set on sills and some fixed directly in the ground without any sill.

Excavations at Greensted in 1960 indicated that the original chancel (and therefore probably also the nave) was of the earth-fast type, for which analogues have been found in Scandinavia.

The work of H.M.Taylor and J.Taylor records evidence of stone churches throughout the Anglo-Saxon period, and as one would expect, there are more surviving stone buildings from the Late Anglo-Saxon Period than from the Early and Intermediate Periods. Apart from the unique wooden church at Greensted and the small amount of literary and archaeological evidence there is nothing to indicate how common wooden churches were and therefore no way of determining the proportion of wooden churches to stone churches at any one time.

The North Door of St. Botolph's Church Hadstock.

The north door of St Botolph's Church, Hadstock (Essex) (see Plate 3:2) was thought to date from c.1020, the date of the consecration, which is believed to have been attended by Cnut. (Jane Geddes, 1982).

The door is made of oak *Quercus sp.* and the construction conveys a great deal of information about the way in which the wood was prepared.

The practice of growing standard trees with coppice encourages the standard to grow tall and straight as it reaches up to the light. The trees were not usually allowed to grow larger than they were needed because a large old tree was likely to be diseased and, once down, caused difficulties in being moved. The tree from which the door at Hadstock was made was very large, as the widest board is thirty seven centimetres wide (14½inches). (author's measurements). As the boards are cut 'on the quarter' (C.A.Hewett, 1982) (see Figs. 3:5 and 3:6) the tree must have been nearly eighty centimetres (31 inches) in diameter when allowance is made for the thickness of the bark. There is a possibility that a tree of this size may have grown in a hedge or field. When a tree is felled the surrounding underwood has to be cleared first, which prevents valuable poles from being damaged. The tree has then to be moved, and in the case of a large tree such as this it is likely that it was cut up on the spot. There is no evidence of saw pits in the late Anglo-Saxon period, and the tree was probably 'cleaved' into planks. (Abbot, 1989). This involves driving wedges into the end so that it splits along the grain. Richard Darrah claims that;

> A large oak tree could take several hours to split in half, however a segment of an oak tree originally 0.9m in diameter and 4m long could be split in half in under two minutes. (Richard Darrah, 1982).

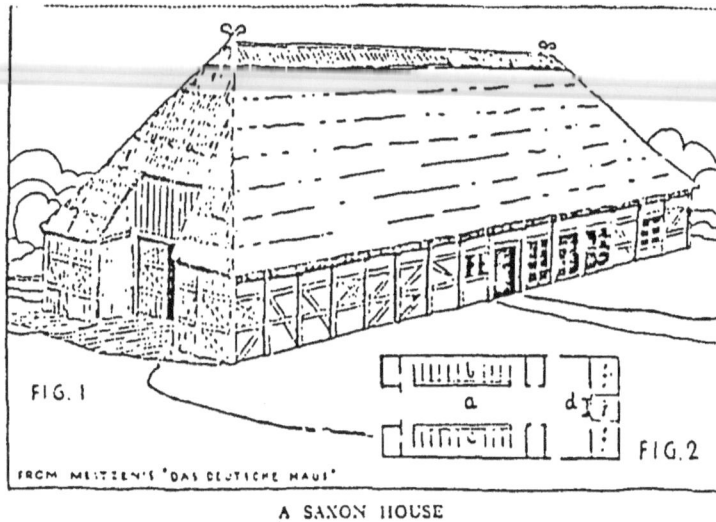

FIG. 1

FIG. 2

FROM MEITZEN'S 'DAS DEUTSCHE HAUS'

A SAXON HOUSE

Figure 3:4   A conjectured Saxon House based on buildings in Friesland in 1882 and reproduced in S.O.Oddy, 1898. This illustration is chosen because it predates the Grubenhausen concept (Leeds, 1913.)

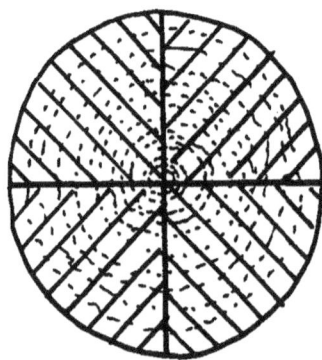

Figure 3:5

Warping caused by cutting across the tree.

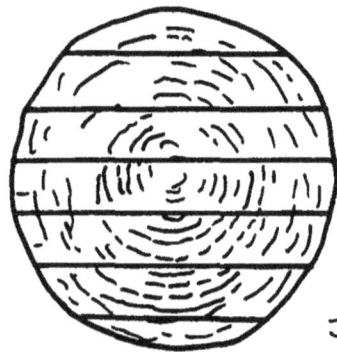

J.B.

Figure 3:6

A quarter cut tree would be split into planks, and then left to season. As Salzman comments;

> Although unseasoned timber was in general use, the leisurely progress of medieval building must have given it time to dry off to some extent. (L.F.Salzman, 1952.)

A tree prepared in this way would season more rapidly and warping would show before the timber was selected for use. The lie of the growth rings in wood cut on the quarter minimises warping, and Cecil Hewett makes the point that the long straight rebates at Hadstock postulate the use of seasoned and quartered oak, and the use of a rebate plane. (C.A.Hewett, 1982). When viewed from the top the ends of the rebates can be seen to still be a snug fit. (Author's observation, September, 1992).

The door is held together by ledges of 'D' shaped section, the most remarkable being the curved ledge which follows the shape of the door. This was bent by heating in steam.

(J.Kelsey, (ed.), 1985). The method involves making a former to the shape of the curve, fastening the former to a solid base and then steaming the wood until it becomes pliable. The strip of wood to be bent is then placed in position round the former. and held in place with pegs and left to become cold. The steaming takes place in a box, and steam is allowed to pass through until the bar to be bent is pliable. This must have tested the ingenuity of the Anglo-Saxon woodworker, although the art of bending wood was practised by shipwrights from the time of the first settlements. (Local tradition in Hadstock attributes the craftsmanship to a ships' carpenter.)

The Hadstock door gives us information on the way in which the standard trees were cut into planks, the size of tree used and the joints and tools which were available to the Anglo-Saxon carpenter. The bending of the ledge shows that sophisticated techniques were available. The survival of the door in such a good state, in spite of the replacement of some sections in the last century, is a silent witness to the skill of one Anglo-Saxon craftsman.

Plate 3:3

The north door of St. Botolph's Church, Hadstock

Part Four

Medicine in Anglo-Saxon Times

In this section we are considering the variety of trees used in Anglo-Saxon medicine. The use of tree products in the Late Anglo-Saxon period cannot be examined without some awareness that some names of plants are not always accurate. As Wilfred Bonser points out:

> Many names of plants in the *Anglo-Saxon Herbal* and in περι διδαξεων though written in Anglo-Saxon are translated from works composed in southern Europe. It will therefore be noticed that many of the plants and animals mentioned are not found in Britain

Although many of the trees mentioned are familiar, there is confusion over some, such as the lotus tree. Bonser also points out that most of the herbs mentioned are no longer used medically. In his preface to the *Leechdoms* Cockayne points out that the treatments are largely influenced by the Greek writers, particularly Hippocrates and, although some leeches were adept at Greek, some of the translation was through the intermediary of Latin. It is not always certain which plant is referred to, as common names were sometimes applied to different species. The confusion of plant names was not resolved until the introduction of the Linnaean system in the eighteenth century.

Cockayne considers the leech to have employed psychology in the treatment, using charms, spells, invocations and astrology, anything in which a patient had faith, to supplement the treatment. The mystique of the doctor has always developed a faith in the medicine and, as we are well aware today, attitude of mind affects the health of the patient

Among the many types of treatment there are some which are no longer used, such as scarification and bleeding. Some however are a familiar part of medicine, although the disease and medicine may differ.

Inhalations were used, the 'gledes' of an aromatic substance being used, such as the pine tree *Pinus sp.* in this treatment for the 'half dead' disease. (Is this stroke? Cockayne offers no explanation.)

> Sume bec lǽreð wið þære healfdeaden adle þ man pintreow bǽrne to gledum and þonne þa gleda sette toforan þam seocum men and þ he wonne ontyndum eagum and opene muþe þane rec swelge þa wrage þe he mǽge. and þonne he ma ne mǽge onwende his neb away lythwon and eft wende to and onfo ðam steme and swa do ǽlce dǽge oð þ se dǽl wǽs lichoman þe wǽr adeadod wǽs and gelewed to þære ǽrran hǽlo becume.

> Some books teach for the half dead disease, that one should burn a pinetree to gledes, and then set the gledes before the sick man, and that he then, with eyes disclosed and open mouth, should swallow the reek, for what time he may; and when he is no longer able, he should turn his face away a little, and again turn it to the *hot* embers, and accept the glow: and to do it every day, till that part of the body which was deadened and injured come again to its former health.

Salves included the blossoms of trees as the salve for spots shows:

> Eft sealf wel on aþydum sceapes smeruwe hægþornes blostman and þa smalan singrenan and wudurofan meng þonne hwitewudu wiþ and hwon buteran.

> Again a salve, boil in pressed sheep's grease, hawthorn's blossoms, and the small stone crop and woodroffe, then mingle mastic therewith and a little butter.

Lotions are prescribed. It is ironic that one for headache is based on willow *Salix spp.*, which is well known as the source of salicic acid, from which acetylsalicic acid is made, which is aspirin. (J.K.Crellin & J.Philpott, 1990).

> Wiþ haefod ece genim sealh and ele do ahsan gewyrc þonne to slyþan do to hymlican and eofor wrotan and þareadan netlan gecnuwa do þonne on þone slipan beþe mið.

> For headache, take willow and oil, reduce to ashes, work to a viscid substance, add to this hemlock and carline and the red nettle, pound them, put them then on the viscid stuff, bathe therewith.

Medicines were concocted with so many items, to be boiled and reduced, that one wonders if it was a case of making sure that at least one thing would produce a cure. For shingles the leech was advised to:

> Genim cwicbeam rinde and æpsan and apuldor mapuldor ellen wiþig sealh wir wice ac slahþorn bircean elebeam gatetreow æsces sceal mærst.

> Take quickbeam rind and aspen and apple tree, maple tree, elder, withy, sallow, myrtle, wich elm, oak, sloe thorn, birch, olive tree, the lotus tree (which Cockayne thinks may be hornbeam), of ash there shall be most.

The recipe continues with instructions to add more herbs, boil and strain and eventually give to the patient to drink.

Salves were made not only from flowers and bark, but from ivy tar which occurs in the salve for wens:

> Wen sealf hiorotes mearh, ifig tearo and gebeaten pipor and scip tearo.

> A wen salve; hart's marrow, ivy tar, and beaten pepper, and ship tar.

Powders were part of the mystique, and the mistletoe *Viscum album* which is mentioned only once, in the cure for shingles, was to be dried, powdered and added to best wine.

> Genim þonne acmistel

> Take mistletoe of the oak.

Fruit soaked in wine was regarded as helpful for a swollen stomach.

> Manigfeald æppelcyn peran æpeninger, pisan of wænda and gesodena on ecede and on wætre and on wine el scearpum

> Manifold kinds of apples, pears, medlars, peas

32

moistened and sodden in vinegar and in water, and in pretty sharp wine.

Twigs were used. In the following case the resins may have been extracted from the upper part of the green twigs of a pine tree *Pinus sp.* which were recommended for liver diseases:

> Eft pinetreowes þa grenan twigu ufeweard gegnid on þ seleste win sele drincan.

> Again rub into the best wine the upper part of the green twigs of a pine tree.Give to drink.

A swollen **Fraenum** or lipstring (T.B.Johnston & J.Whillis, 1954) was treated with the kernel of a peach *Persicum malum*.

> cyrnel þ byð innan þan persogge.

Fomentations were made from bunches of laurel, *Laurus sp.* (berries or flowers) which were advised for a liver abcess:

> On þam wætre sien gesodene laures croppan

> On the water let there be sodden bunches of laurel. In this case it was as part of a hot fomentation.

Emetics are ways of inducing vomiting and, as well as tickling the back of the throat with a feather or fingers, several potions are suggested. This one uses elder *Sambuscus sp.* bark among the ingredients:

> Spiwe drenc hamwyrte III snæda and ellen rinde berende gelice micel XXV lybcorna gegnid do huniges swilce an snæd sie on ete þonne mid cuclere on sup hates wæteres oððe cealder.

> Spew drink of home wort three pieces, and rend up elder rind, the same quantity, twenty five libcorns, rub them to dust, and of honey as much as would be one piece, then eat there of with a spoon sip some water hot or cold.

Purgatives are still to be found in herbals (Gaea & Shandor Weiss, 1985) and include the bark of elder *Sambucus sp.*:

> Ellenrinde neoðwearde.

> Elder bark from the lower part of the tree.

Dentistry was in demand and there are several recipes for the alleviation of toothache. The one cited here gets to the root of the trouble, perhaps indicating that the patient, then as now, will accept great discomfort in the hope of relief:

> Wiþ toðwærce hnutbeames rinde and þorn rinde gecnua to duste adrig on pannan snið utan þa tew scead on gelome.

> For tooth wark, bray together dust rind of nut tree and thornrind, dry them in a pan, cut the teeth (gums) on the outside, shed on frequently.

Holly *ilex aquifollum* is the basis of another leechdom for toothache:

> Gif wyrm ete þa teð genim ofer geare holen rinde and eofor þrotan moran wel on swa hatum hafa on

muþe swa hat swa þu hatest mæge.

> If a worm eat the teeth, take holly rind over a year old, and root of carline thistle, boil in so hot water hold in the mouth as hot as thou hottest may.

Mouthwashes were used and this indicates that oral hygiene must have been a problem, although tooth picks are mentioned. A mouthwash for sores was prescribed.

> Wiþ innan tobrocenum muðe nim plum treowes leaf wyl on win and swile mið þone muþ.

> For a mouth troubled with eruption within; take leaves of the plum tree, boil in wine, and swill the mouth therewith.

Deafness must have been a recurrent problem and this treatment using alder *Alnus sp.* berry juice does not sound very promising, but it may be an attempt to soften wax.

> Wiþ earena adeafunge eft ellencroppan getrifulad þ seaw wring on þ eare.

> For deafening of the ears again, alder bunches triturated, wring out the juice into the ear.

Sore Eyes were treated kindly, with ointment or drops, as this simple treatment shows.

> wiþ eagece genim wiþowindan twigu gecnuwa awylle on butern do on þa eagen.

> For eye ache, take some twigs of withewind, pound them, boil them in butter, apply to the eyes.

Cancer must have been a problem for the doctor, and for this disease leaves of the nut *Corylus sp.* tree and oak *Quercus sp.* were used with a great many herbs in a salve.

> Hnutbeams leaf *and* ac leaf.

> Leaves of the nut tree and the oak tree

(This would be the hazel nut.)

Heart Disease was another complaint in which the doctor felt incompetent. The treatment for heart disease was blackberry *Morum sp.* juice:

> Drinc seoca of bræmel berian gewrungene eft

The sick *man* drink of bramble berries wrung out often

It is not possible to determine how accurately the diagnosis of a disease conforms to our modern knowledge. Cancer, for instance, is likely to have been limited to growths showing on the outside of the body, and the term 'leprosy' was probably applied to many skin complaints. The Anglo-Saxon physician had many medicines available in the woods, and this emphasises the reliance of the Anglo-Saxons on the products of the woodlands.

Dyes in Late Anglo-Saxon England

Dyes have been known for many thousands of years, but because of the nature of fabrics relatively few specimens survive. The use of dyes is largely inferential, but Gale Owen-Crocker makes the point that;

The colourful nature of their dress is constant evidence of their exuberance—- the love of colour indicated by the bright shades of the manuscripts and needlework confirm it.

The merchant in Ælfric's Colloquy states that he imports;

> pællas *and* sidan, deorwyrþe gymmas *and* gold, selcuþe reaf *and* wyrtgemange, win *and* ele.

> purple and silk, precious gems and gold, various garments and pigments, wine and oil. (Garmonsway, 1939)

Purple may refer to cloth or dye, and the word wyrtge-mange glosses pigmenta but in Garmonsway's glossary it is described as 'mixture of herbs, spices, perfume' which does not agree with the Latin. The author suggests that in this case a more accurate translation may be vegetable dyes.

We know that the Anglo-Saxons used dyes from the words *dæg*, *wad* and *madder*.

Dæg is used in Ælfric's Catholic Homilies

Se wolen-reade wæfels wislice getacnode ures Drihtnes deað mid dære deage hiwe

The scarlet robe wisely betokened our Lord's death by its dyed colour.(Thorpe 1846)

Woad *Isatis tinctoria* is a plant which gives a blue dye but does not require any extract from trees. The reeve in 'Gerefa' is instructed to maderan setta 'set madder', in the Spring.(Liebermann, 1903) Madder *Rubia spp.* root was the source of one of the principal dyes. Madder dye can only be obtained from the root of the madder plant, where it occurs in an acidic form, (Mairet, 1917) by the action of lye,

extracted from wood ash.(Robertson, 1969, Battiscombe, 1956)
Unfortunately the silks were probably imported ready dyed and so cannot be regarded as evidence for the use of these dyes by the Anglo-Saxons. Wool was used in the embroidery of the Bayeux Tapestry, and this is probably the best source for analysing the dyes. Destructive analysis methods have been the only way of determining the dyes used in the past, but the author suggests that spectro-analysis, which uses reflected light, may be available in.the future

The nature of dyes is fairly complex, there being two groups, substantive and adjective. A substantive dye imparts a colour without any additional preparation, while an adjective dye requires the use of a mordant, applied to the fabric, generally before, but occasionally with or after the dye. Grierson, 1989)The most usual mordant is alum, which is a double salt of aluminium sulphate and iron sulphate.(Singer,1948) It occurs in the volcanic area near Naples, and was known throughout the ancient world (Singer, 1948)· It is referred to by Pliny, who notes that it is also an astringent.(Bailey, 1929-32)· Iron sulphate, otherwise known as copperas or green vitriol, was also used.(Fowler & Fowler,1950)

The fabric, wool or silk, has to be cleansed by boiling with soft soap and rinsing well, otherwise the mordant is unevenly distributed, causing an unevenness in the dyeing.(Mairet, 1917) Alum is the essential mordant for madder (Singer, 1948) and, from the evidence of the vocabulary for the use of madder, alum may well have been imported in tenth century England.

Although the earliest record of alum imports is of the twelfth century, the Old English word 'efne' which glosses the Latin 'Alumen, vel stipteria' in the *Nominum Herbarium* suggests that alum was known in Late Anglo-Saxon England. While it is not possible to be certain which dyes were used in Anglo-Saxon England we can list the plants and colours available. Appendix 2, below, based on E.M.Mairet's *A Book of Vegetable Dyes* shows what was available from trees. Other colours were available from other plants, to give a very wide palette.

Part Five

The low temperature processes using wood, wood ash, lye and bark.

Wood as Fuel for Anglo-Saxon Industry

As well as domestic fires, wood was essential for industry, for salt pans and lime kilns, for glass making and as charcoal for smelting metals and smithying.

The use of wood for salt furnaces is referred to in a charter of Bentley, near Holt, Worcestershire. Bishop Oswald in the tenth century (962) (S.1301) required a supply of wood for salt production.

> ad coctionem salis —— et silvam necessariam on Bradanlæge ad illam præparationem salis
> .
> For the cooking of salt —— and the woodland necessary for the preparation of salt. (Hooke, 1981).

Domesday Book records that the king's manor of Bromsgrove received;

> salinae CCC mittas salis, quibus dabantur T.R.E. CCC caratades lignorum a custodibus silvae.
>
> 300 mitts of salt, for which they used to be given 300 cartloads of wood by the keepers of the wood in the time of King Edward. (Hooke, 1981).

An apparently unrecorded use of wood was the large scale wood burning for the production of lime for plaster and mortar, used in the walls of the surviving churches described by H.M.Taylor and J.Taylor in *Anglo-Saxon Architecture.*

The Uses of Wood Ash

The ultimate end of a great deal of wood in Anglo-Saxon times was probably firewood. The consumption of wood on household fires, for cooking and warmth, created a constant demand, and various industries used wood, as logs or as charcoal for fuel. We are also aware of the use of wood for heating salt pans and of charcoal for smithies and metal smelting. This resulted in the production of wood ash, which was not wasted. Mixed with water it gave lye, and on its own was used in the glass industry. The glass industry used soda ash up to the ninth century, when potash replaced it. (R J.Charleston, 1991) This gives an inferior glass which tends to devitrify. (D. B. Harden, 1956) This is to be found throughout Northern Europe in the ninth and tenth centuries. (R.J.Charleston, 1991).

The evidence of the glass suggests that supplies of natron,

probably from Egypt, ceased, and the replacement with potash, from wood ash, was a natural alternative. The evidence of soap making, which also used wood ash, is consistent with this, as it is thought that hard soap, made from soda ash, was known in the Arabic world from the first century. (F. Sherwood Taylor, 1957). Wood ash has potash as a major constituent, which gives soft soap

Soap in Anglo-Saxon Times

The nature of soap prevents it occurring in the archaeological record, but it is referred to in documents both directly and by inference. The early ninth century Continental document, *Capitulare de Villis vel Curtis Imperialibus* refers to:

> (item 24) cleanliness and quality of food provided for the table, and
> (item 34) lard, smoked meat, sausage, newly salted meat, wine, vinegar, mulberry wine, boiled wine, garum, mustard, cheese, butter, malt, beer, meads, honey, wax and flour are to be prepared with the greatest attention to cleanliness. (H.R.Loyn & J.Percival (eds.) 1975).

The concept of cleanliness implies the use of cleansing agents and it therefore comes as no surprise to find that the women's workshops are to be supplied with, among other things, soap and oil, (item 43) and, in (item 44), soap is to be part of the rent. The steward was also expected to give:

> Three pounds of wax and eight *sextaria* of soap each day he is in service. (item 59)

Item 62 requires that an annual statement be made of all income, including that from 'tallow and soap'. (H.R.Loyn & J.Percival, 1975).

In England the concept of bodily odours led to the *Leechdoms* recording:

> Ðeos wyrt ðe man scolmbos nemned on wine geþylled and geonþuncen heo þone fulan stenc ðæra oxna and eales þæs lichaman asyrrew
>
> This wort, which is named σχολυμος boiled in wine, removeth the foul stench of the armpits, and of all the body. (Cockayne,1864-66)

Although the efficacy has not been tested the need for a deodorant was clearly understood, and an awareness of bad breath is reflected in the instruction in the *Leechdoms*:

> if monnes oð sie ful genim beren mel goð and clæne hunig and hit sealt gemeng eall tosomme and gnid þa teþ mid swiðe and gelome
>
> for foul breath take barley meal and clean honey and white salt, mingle altogether, and rub the teeth with it much and frequently. (Cockayne, 1864-66)

Theophilus, writing in Northern Germany, refers in his treatise *De Diversus Artibus*, to soap as a well accepted common substance. As his work describes how to make or prepare various artifacts it would seem that soap was well known at the time of writing, which has been variously dated between the ninth and eleventh centuries.Soap is mentioned with hog's lard, and with bristle brushes. (Dodwell, 1961).

In the introduction to his edition of Theophilus, Dodwell refers to the *Lumen Animae* as commenting on the manufacture of soap and cosmetics, but as it is of a much later date it has not been included as a point of reference, but is regarded as confirming what Theophilus himself speaks of as commonplace. (Dodwell, 1961).

The *'Plan of St.Gall'* refers to soap and oil being issued to the monks with their clothing;

'saponem sufficienter et uncturam'

If soap was a common item in Carolingian times then it had probably reached England by the end of the tenth century. The soap maker's trade was centred on Bristol by 1180, (Sherwood Taylor, 1957), and it seems that it was made from tallow and ashes. We should not however think of soap as the scented solid with which we are familiar today. It is necessary to consider the chemistry and the likely source of the components which were available.

It is important to realise that there are two types of soap, soft soap and hard soap. The basis of hard soap is sodium, and soft soap is based on potassium. The lye which is made from wood ash in Northern Europe is potassium based and the wood ash from Mediterranean countries with a high salt content in the soil has more sodium in comparison. This was further complicated by the fact that a process known as 'salting out' was introduced in the seventeenth century. This introduced sodium, making soap harder. Soap made in England before this time would have been soft.

The chemical reaction which takes place in the manufacture of soap is that of a strong alkali reacting with an organic acid, generally caustic soda with stearic acid (which is the main constituent of animal and vegetable fats). In chemical terms the reaction is represented as:-

caustic soda + stearic acid = sodium stearate + water
$$NaOH + \text{stearic acid} = Na \text{ stearate} + H_2O$$
(Chemical symbols are not used for organic compounds in order to keep the explanation as simple as possible.

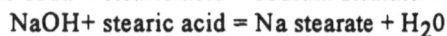

In the absence of caustic soda, caustic potash may be used and this gives a soft soap. It is suggested that this may have been referred to by Pliny when he used the word *sapo* to describe a pomade invented by the Gauls and is referred to by Priscian c.385 A.D. as a shampoo. It seems to have been a common domestic craft by c.800 in Europe. The possibility of caustic soda being used is low, as the principal source

of sodium carbonate was Egypt, where it occurs as natron, or the wood ash from Mediterranean lands where it occurs in halophytic plants. Wood ash was easily obtained in Northern Europe but here it contains potassium carbonate. The ash was used to prepare lye, from which soap was made. An alternative source of sodium is salt (sodium chloride). There are ample supplies of salt in England, and there was a thriving salt industry in Anglo-Saxon times (D.H.Hill, 1981) but it was not until the seventeenth century that salt was used in the manufacture of soap.(Sherwood Taylor, 1957) This provided sodium which is the vital element required for hard soap.

We must therefore consider the traditional way to make lye, in order to understand the nature of the raw material. As every one who has cooked over a wood fire knows, the best way to clean a greasy pan is to use a handful of wood ash and a small amount of water. The fat combines with the potash in the ashes to form a type of soap, which is then easily swilled away. This is the basis of the manufacture of soap. The traditional farmhouse method, used up to the beginning of this century, was to take an old barrel, bore holes in the bottom, put a layer of gravel and then some straw to help with drainage, and then fill the barrel with wood ash. The ash from hardwood was better than that from softwood, (Sherwood Taylor, 1957) since softwood grows more rapidly and so does not accumulate as much sodium and potassium as the slower growing hardwoods. Rainwater was then percolated through the ashes and the liquid collected at the bottom was then concentrated by boiling it down until a fresh egg would float in it. (Sherwood Taylor, 1957).

Lye was used as a caustic solution in the making of leather and soap and is mentioned by Theophilus as a flux for gold soldering (beechwood ash is specified), (Dodwell, 1961) and in the extraction of dye from madder. (Dodwell, 1961) To make soap the specific gravity was adjusted by taking a saturated solution of salt, floating a weighted stick in it (so that it floated upright) and using it as a hydrometer in the lye. (Seymour, 1984) In this way a consistent strength was achieved and so the same proportions of fat and lye could be used each time. Fat can be measured by weight or volume and the lye by volume.The author has made soap from wood ash in a manner which is probably similar to that which would have been used in the Late Anglo-Saxon Period, details of which are in the Appendix. Clothes were cleaned in antiquity with fuller's earth (hydrated aluminium silicate), and alkali, often urine which had been left to stand and become ammoniacal (Sherwood Taylor, 1957) The fuller's earth absorbs fat and alkali forms soluble compounds with fat.

Lye was 'sharpened' by the addition of lime, which increased the alkalinity. Sherwood Taylor is of the opinion that this was done from antiquity. (Sherwood Taylor, 1957) Here the chemistry becomes more complex as the molecules in solution act as strong alkalis and weak acids, so that the alkaline properties predominate.

potassium carbonate + water = potassium oxide + carbonic acid

$$KCO_3 + H_2O = KO + H_2CO_3$$

calcium carbonate (lime) + water = calcium oxide (quick lime) + carbonic acid

$$CaCO_3 + H_2O = CaO + H_2CO_3$$

At this point we can consider the advice given in the *Leechdoms*, especially the treatment for scabies, which runs;

> Genim gose smero and niþewearde elenan and haran sprecel, bisceopwyrt and hegrifan; þa feower wyrta cnuwa tosomne wel, awring, do þæron ealdre sapan cucler fulne; gif þu hæbbe lytel eles, meng wiþ swiþe and on niht alypre. (Cockayne, 1864-66)

> Take goose-grease and the nether end of ele-campane, and viper's bugloss, bishopswort and hairif; pound the four herbs together, squeeze them out add thereto a spoonful of old soap; if you have a little oil, mingle it thoroughly (with the foregoing], and at night lather (the mixture] on.

A soft soap would mix with oil, whereas hard soap would need grating and making into a solution first. The act of lathering must have been well known, and indeed the Bayeux Tapestry shows clean shaven men. (Musset, 1989). It is not easy to shave without soap of some sort to soften the beard.

There are other references to soap in the *Leechdoms,* where old soap and marrow soap are particularly preferred. (Cockayne, 1864-66). Old soap has matured and is less likely to be caustic, as the excess unstable hydroxyl radical will convert to water by releasing oxygen.

$$(OH)_2 = H_2O + O_2$$

Soap was probably also used to wash wool before dyeing; if it was not washed the mordant would not take evenly. (see above.)

Soap making only required gentle heat, fat and wood ash. A pot on a dying fire would maintain a constant supply of soap.

Leather in the Late Anglo-Saxon Period

The process of leather making requires wood ash to make lye for removing fat from the hide and bark from the oak tree to impregnate the hide with tannin.

Bosworth and Toller state that the word 'leþer' is invariably found in combination with another element, such as leþer-wyhrta (Leather worker). This is well illustrated in Ælfric's *Colloquy;*

> Ic bicga hyda *and* fell, *and* gearkie hig mid cræft minon and wyrce of him gescy mistlices cynnes,swyftlera *and* sceos, leþerhosa *and* buteri-cas, bridelþwancgas *and* geræda,flaxan *vel* pin-nan *and* higfatu, spurleþera *and* hælftra, pusan *and* fætelsa; *and* nan eower nele oferwintran buton minon cræft.

> I buy hides and skins and prepare them by my craft, and make of them boots of various kinds, ankle leathers, shoes, leather breeches, bottles, bridle thongs, flasks and budgets, leather neck pieces, spur-leathers, halters, bags and pouches, and nobody would wish to go through winter without my craft.

(A water budget consisted of two bottle shaped limp bags carried by their necks on a t shaped pole.) (Singer, 1956) (See Figure 5:1)The list given by Ælfric in his *Colloquy* gives gives us some idea of the uses to which leather was put.It was also used to bind books, such as the Stoneyhurst Gospel. (See Figure 5:3)

The archaeological record is scanty, as it is with most organic objects, but the anaerobic conditions of the York excavations have revealed shoes, laces, belts, garments, bags, sheaths and gloves. There is a clear illustration of a pair of boots in the Canterbury Psalter. (See Figure 5:2)

There are three basic processes for the making of leather.

(a) The oil process or chamoising, which consists of cleaning and drying the skin,and then rubbing in oils and greases, which enter the fibres and oxidise, making the leather very soft, supple and strong. In spite of its name, chamoising of leather is usually applied to sheepskin.

(b) The mineral process, which uses alum to dry and preserve the skin, which is then softened by 'tawing' (pulling and bending over a curved blunt blade). This was important in the middle ages, but is rarely used now.

(c) The vegetable process, or tanning. This used two products from the woodlands in the Late Anglo-Saxon Period, lye and oak bark, and is the one in general use today.(Singer, 1956)

It cannot be proved or disproved as to whether or not method (a) was used, but the likelihood of method (b) is slender. The word alum (O.E., efne) occurs only once (as a gloss in Wright's Vocabulary) (see also the section on dyes.) In a major industry like tanning it is likely that the word would be more widely used.(Although it must have been widely used in the use of madder)

The existence of tanning pits in the archaeological record indicates the the use of method (c).

The purpose of tanning leather is to make the hide into a supple preserved skin and this is basically achieved by removing hair and subcutaneous fat and then impregnating the tissue fibres with tannin. At a later stage the tannin combines with collagen, the protein of which the fibres are composed.(Blair & Ramsey, 1991).

The methods changed very little until the nineteenth century, (G.W.Humphreys, 1966.) and the old ways can still be seen in less developed countries, such as Morocco and Egypt. Oak bark was taken from trees felled in March or April,when the sap was rising.(Arnold,1970) The bark was stripped off the tree and shredded before being put into the vats. The tannin is in the bast, immediately under the the actual bark, but in practice it is not practical to separate the two. Tannin occurs in most trees, and the bark of the larch tree is the major source today

In (c) the hide is washed to remove blood and dirt. The hide is trimmed and the feet and horns which are still attached when the butcher has finished with the beast are removed. The presence of these bones and horn cores in archaeological excavations are one of the clues that a tannery was present, which is complemented by the discovery of tanning pits, such as those at Tanner Street, Winchester.(Blair & Ramsey, 1991)

The hide is folded, hair side in, and left until the rotting process has loosened the hair.(Blair & Ramsey, 1991) The hair is scraped off and the hide is then soaked in lye or peur to loosen the fat. Peur is a solution of dog dirt, used warm, which contains the bacteria required to break down the fat, An alternative mixture was poultry droppings (used cold). (Singer, 1956). Lime water was used in the Middle Ages to loosen hair, and had the advantage of 'plumping' the skin, making the impregnation of tannin easier. Peur removed lime and dissolved albumin. The plumping by lime made the leather flabby, and suitable for clothing, purses, bags etc,(Singer, 1956) The date at which the use of lime started is uncertain, although D.M.Wilson in *The Archaeology of Anglo-Saxon England* suggests that carboniferous limestone found near tanning pits at York may have been a source of lime in the slaking process.(Wilson, 1976).

After the hair and fat was removed the hide was soaked in the tanning pits, in solutions of increasing strengths. The first pit contained an old and mellow solution,which had been used before. The gradual increase in strength was to allow the thorough impregnation of the tannin, a strong solution too soon would prevent subsequent immersion from penetrating to the inside of the hide. The soaking and draining process could last for a year or fifteen months, and sometimes finished with the hide being folded round oak bark.(Blair & Ramsey,1991) The final process, carried out by curriers, was the rubbing in of a dubbin made from a mixture of tallow and fish oil to give the leather suppleness.(Blair & Ramsey, 1991)

The vast quantities of bark required for tanning hides is indicated by Edlin (1973) in *Woodland Crafts of Britain,* who tells us that it has been estimated that ten hundredweight of bark, which yields one hundredweight of tannin, was needed to treat two hundredweight of fresh skins. The debarked trees could be used to make charcoal or as timber for buildings or smaller poles for fences, but not tool handles, for which use oak is unsuitable. The demand for the products of the whole tree would have made full use of the resources of the woodlands.

Street names such as Scyldwyrhtana Stræte at Winchester in 996 (S 889) (believed to be in existence by 990), (M.Biddle, 1964) indicate an established leather industry and so a constant demand for wood ash for lye and oak bark for tannin.

Figure 5:1
Water budget carved on the font of Hook Norton
Church, Oxfordshire.

Figure 5:2
Detail from the Canterbury Psalter, folio 195b.

Figure 5:3
The Stoneyhurst Gospel is covered with red leather,
with intricate tooling. It is on wooden boards, which
together with the penmanship on vellum, is a good
example of Anglo-Saxon craftsmanship.

Part Six
Charcoal in Anglo-Saxon Times

The use of charcoal in Anglo-Saxon times has already been mentioned (Rackham, 1976) and once again the evidence is inferential, as charcoal as a prepared fuel does not appear in the archaeological record, although its appearance in excavations, as evidence of hearths or accidental ignition of wooden structures is well documented.

The use of the word col is noted by Bosworth and Toller as including the simile 'as black as coal' *swa sweart swa col* (Bosworth & Toller, 1838-1921) and in the *Leechdoms* the phrase on hat col 'upon a hot coal' is used in reference to the use of coal to cauterize the foot infected with worms. (Cockayne, 1864-66)

Place-name evidence is sufficient to confirm that charcoal was widely used, and the name element *col* is present in Coldred, Kent, (Colret in 1086 (DB) 'a place where coal is found or charcoal is made')and Coldridge, Devon (Colrige in 1086 (DB) 'ridge where charcoal is made'.Transport of fuel is indicated in Coleford, Glos. (Coleforde in 1282) 'ford across which coal is carried', and Coleford, Somerset, near Frome (Culeford in 1234) 'ford across which coal or charcoal is carried'. Colerne, Wilts (Colerne in 1086(DB) means 'a building where charcoal is made or stored'. (Mills, 1991).

Apart from the alternative meaning of coal, where the local geology can help in determining the exact meaning (the Kent coalfield clouds the issue in the case of Coldred), there are other explanations for 'col' in modern spellings of a place-name. It may be from a personal name; or from an Old English word meaning 'cool'; or it might be from a Celtic river name. Such alternatives occur in Coleby, Humberside; Colebrook, Devon; and Coleshill, Warwickshire (Colleshylle in 799). The etymology must therefore be traced back to the first reference for one to be reasonably sure of the interpretation.

The Scientific basis of charcoal

The manufacture of charcoal is basically one of wood distillation or the incomplete burning of wood. The wood, of fairly uniform diameter, is stacked in a pile and a hole left in the centre in which a fire is lit, as shown in Plate 6:1. The stack of wood is covered with turf and a few holes left to admit a limited amount of air. The wood smoulders and the charcoal maker watches the 'clamp' carefully, removing or replacing turves so that the stack does not burst into flame or the 'fire' go out. This process may continue for as long as two weeks, and then has to cool gradually. Premature opening of the clamp will result in the stack igniting, and so result in the loss of the whole batch. The author has witnessed this operation in Zambia, where the bush is being cleared for agriculture and the charcoal used as fuel for domestic cooking and by the copper mines in the smelting process.

Ure's Dictionary gives an account of charcoal making and gives an analysis of peat charcoal as; charcoal 30%, tarry products 3%, watery products 30%, gases 37%, (R. Hunt (ed) 1860) There is also a description of the collection of liquid condensates by having the clamp on sloping ground, and sometimes putting pipes to collect the liquor.[It is assumed that these watery products are terpenes, sugars and condensed saps]. A more recent account is given in the *New Scientist* article 'Plant a Tree for Chemistry.' (John Emsley, October, 1987.)

Charcoal gives a smokeless fire with an even heat, which makes it popular for barbecues. Its industrial uses are as a reducing agent for the extraction of metals from their ores and, used with a blast of air, very high temperatures can be reached. In Anglo-Saxon England the smiths must have used charcoal extensively. Forged weapons, such as swords, became steel (steel is iron containing a small percentage of carbon) and in processes with other metals the reducing atmosphere prevented oxidation.

The more affluent classes probably used charcoal as a cooking fuel, and its use as a smokeless fuel in draught-free halls may have led to unexplained deaths due to carbon monoxide poisoning. (Zambian radio stations frequently broadcast warnings about burning charcoal indoors during the winter when occasional frosts occur.)

Tar as a by product of charcoal

The chemistry of wood is fairly complex, but basically it is cellulose containing other materials, such as resins and turpenes, some of which burn during the preparation of charcoal. However, because the best charcoal is produced by burning at low temperatures, not all of these materials are lost. Conversation with a professional charcoal burner revealed that before the modern use of kilns a good charcoal maker could get a layer of tar under the turves in a charcoal clamp. This could be torn away in sheets. On a subsequent visit a sample of this tar was obtained. This by-product was almost certainly used in Anglo-Saxon times as the word teoru(o), used in the *Leechdoms* indicates;

> scip tearo and garlic
> ship tar and garlic.

> scip tearo and sceapen smeara
> ship tar and sheep grease

The recipes for salves include tar.

> Wen sealf hiorotes meorh, ifig tearo and gebeaten pipor and scip tearo

> Wen salve harts marrow, ivy tar and beaten pepper and ship tar.

Meng þonne wið piper *and* wið tearon gegrind smer mid.

mingle with pepper and with tar, grind these, smear therewith.

Wið hondwyrme nim sciptearo and spefl

Against a handworm take ship tar and sulphur

'Reference to tar also occurs in *Beowulf;* (line 295) (M.Swanton, (1978)

niwtyrwynde nacan on sande

The newly tarred vessel on the sand.

The etymology of the word provides more evidence that tar was obtained from trees, as *The New English Dictionary on Historical Principles says*:

Tar, teoru(o) Generally considered to be a derivative of O.Teut. *trewo* Goth *triu* O.E. *trewo* (Indo-Eur, *derw-*: *dorw-*: *dru*) c.f. Lith dorvia (pinewood), Lett. *darwa* (tar) O.N. tyr- *vid* (pinewood) Thus *trewo* may have meant originally 'the product (pitch) of certain kinds of trees'.(Murray & Bradley, 1909)

Tar occurs in the archaeological record in the excavation of the Graveney boat, where it was used in caulking. Samples Gr255, 264 and 265 were identified as wood tar by paper chromatography.(Fenwick, 1978)

A similar substance to tar is pitch which also occurs in the Anglo-Saxon vocabulary. The word 'pican', appears in the *Leechdoms*;

gedo on water xxx nihta on anne crocken þ on þe sie ge piced utan
put into water for thirty days in a crock, one that is pitched on the outside. (Cockayne, (1864-66).

The type of pitch is also specified at one point

do to þam ele clane buteran pund hlutnes picer fiftan healfyntan.

add to the oil a pound of clear butter and a half of clear pitch.

wiþ þon ilcan eft beran melo and hluttor pic

Mingle together beer meal and clear pitch.

The New English Dictionary also gives a quotation from the Comp Farmer of 1766.(A publication it has not been possible to trace,although there was a *Complete English farmer* in 1771).

The branches resemble those of pitch trees, commonly called the spruce fir, (Murray & Bradley, 1888-1928)

The other product of trees is the sap, which is made up of various substances according to the species. The most well known of these are latex from the rubber tree and maple tree syrup. In the tenth century Theophilus described the use of sandrac, (Dodwell,1961) which comes from the tree *Thuya articulata*, which is a conifer native to North Africa. (Hunt (ed.),1860).He also describes how to use the gum from the cherry or plum tree as a drying agent in varnishes. (Dodwell,1961) From this information we may be sure that exudates from trees were used in medicine, boatbuilding and painting.

Plate 6:1

Section of a charcoal clamp at Hey Bridge,
Bouth, Ulverston

Charcoal is prepared by burning the wood slowly by controlling the supply of air with the turves which cover the clamp. This is a skilled job and the charcoal burner lives on the spot during the charcoal burning season. A 'bothy' is built of branches, sometimes with a dry stone wall if there is a supply nearby. It is necessary to be present during the whole time of the burn; too much air will result in the clamp bursting into flame and the charcoal is lost, too little air and the fire will go out. Although the word 'bothy' does not appear in Old English it appears to be related to ' booth' which has Scandinavian origins.(Fowler and Fowler, 1964)

Part Seven
High Temperature Processes

Glass production
Smelting of lead and iron

Glass in Late Anglo-Saxon Times

Glass was not widely used in Anglo-Saxon times, and was certainly regarded as a luxury, as the paucity in the archaeological record shows. The word glæs occurs in Bosworth and Toller's dictionary, and in Ælfric's *Colloquy* it is included in the list of goods which the merchant imports:

Tin, swefel and glæs, *and* þulces fela.

Tin, sulphur and glass and many such (things). (Garmonsway, 1939)

The word is also used in Ælfric's *Homilies,* in the account of the deposition of St. Martin, Bishop:

His lic weard gesewen sona on woldre, beorhtre ðon glæs, hwitre donne meoloc

His corpse forthwith appeared in glory, brighter than glass, whiter than milk. (Thorpe, 1846)

Ðæt scire glæs

'the clear glass'

appears in the poem 'On The Day of Judgment in Thorpe's *Exeter Book* (Thorpe, 1846) and in Cockayne's publication of 'King Alfred's Book of Martyrs' in *The Shrine* we have

Hi toslogon his glæsenne calic

They broke his glass chalice. (Cockayne, 1864-70)

There is no place-name evidence for the manufacture of glass in pre-Conquest England, he sole entry in Mills' *English Place-names* being Glascote, Staffs. Glascote 12th century, 'Hut where glass is made', OE glæs + cot, (Mills, 1991).

Although glass making has been practised from antiquity, (A.Neuburger, 1930) and was used in Britain in Roman times, the art was not carried on by the Anglo-Saxons and its re-introduction seems to have been difficult. Wilfrid, Bishop of York, had glass inserted in windows of King Edwin's church of St,Peter when restoring it in 669, or more probably 678. Previously they had been fitted with a fretted slab (Harden, 1961) The appeal to Gaul by Benedict Biscop for glass makers in 675 was followed in 758 by Cuthbert sending abroad again. This suggests that the industry had not become established. (Harden, 1961)

Archaeological evidence however suggests that by the Late Anglo-Saxon period the craft had a foothold and excavations at Glastonbury Abbey by C.A.Ralegh Radlord indicate that a large glass house was operative in the ninth to tenth centuries. (Harden, 1961) Fragments show that vessels and window glass were being made. The bubbles are elongate in parallel lines, which is typical of cylinder blown glass. D.B.Harden suggests that the large area on which the glasshouse stands, with three low temperature furnaces,

indicates enough glass production to supply glass for other buildings in the vicinity. (Harden, 1961) Window glass has also been found at Kingsbury, near Old Windsor, Berks, where Hope-Taylor found fragments of thin window glass and two pieces of 'H' shaped calmes in the foundation fillings of the Late Saxon Hall. (Harden, 1961). At Southhampton Maitland Muller discovered fragments of cylinder blown glass in 1947/8, but it must be remembered that Southampton was a port during the eighth to tenth centuries, so these could be of imported glass. Professor Moore has identified a piece of cylinder blown window glass from Thetford. (Harden, 1961). Harden also comments that at the time of writing (1961) reports of these finds had not been published) The fragments of glass found reflect the poor quality, and the possibility that remnants may go unrecognised due to the devitrification process. Glass is an extremely complex substance, and from pre-Roman times to c.1500 it was made from sand and plant ash. This was heated until molten glass was formed of sodium or potassium silicate, at temperatures of over $1000^{\circ}C$, depending on the composition.

R.J. Charleston makes an interesting observation in his article 'Vessel Glass' in *English Medieval Industries:*

Up to the beginning of our period soda ash seems to have been available in Northern Europe, but after about 1000 these supplies seem to have been disrupted or to have dried up, and the northern glassmakers were forced back on the ashes of inland vegetation. (Charleston, 1991).

Charleston also observes that these supplies are not only rich in potash but also have a reasonably large lime content, which gives stability to glass. Soda ash would probably be imported from the Middle East, where the salty soil supports halophytic plants. (If the importation of ashes sounds unlikely it should be borne in mind that by 1500 wood ash was being imported from the Baltic countries for the soap industry.) It is possible that natron, the mineral form of sodium carbonate, was used instead of ashes, for Pliny recognised that nitrum, the old name for natron, was used for glass making and mummification. (K.C.Bailey, 1929-32). It was not until the sixteenth century that Agricola recommended the addition of 'salt from brackish or sea water', which would add the element sodium to the mix, the lack of which we now know caused the devitrification of the potash glass. (Harden, 1961).

Wilson also points out that the use of soda glass ceases in the ninth to tenth centuries. (D.M.Wilson, 1976). The use of potash produced an inferior glass which tended to devitrify, and normally weathered far more quickly than soda glass, developing an enamel-like surface in spots which when occurring on both sides of the glass caused the glass to disintegrate. (Harden, 1971)

The technique used for window glass production was the tubular or muff method, described by Theophilus.(Dodwell, 1961) The glass is blown into a 'bladder ' shape, and then cut lengthwise and opened up to give a sheet of glass. (The better known method of spinning a disc (the 'crown method') originated in fourth century Syria but was not developed in the west until the fourteenth century when a centre was established near Rouen. (D.B.Harden, 1971) Theophilus recommends the use of beech ashes. (Dodwell, 1961) [We now know that beech has a high proportion of magnesium which makes a clearer glass.] (Harden, 1961) This is mixed with sand, and the presence of iron oxide in the sand gives a green colour, present in nearly all medieval glass.(Charleston,1991). The shape of the pot in which the glass was melted was described by Theophilus, and the foot of a pot and rim fragment from the pre-Conquest Glastonbury site have the features of an out turned lip and an everted side.(Charleston, 1991).

Theophilus also describes the making of the glass furnace, which has two compartments, and another furnace for the annealing process. A third furnace was for spreading and flattening the glass. (Charleston, 1991). Theophilus understood the process of making a frit from ashes and sand, (Dodwell 1991) but does not mention the addition of lime, although he recommends it when annealing painted glass. (Dodwell, 1961). The potash or soda ash acts as a flux, causing the silica to melt at a lower temperature.
Potash/lead/silica glass melts at c.1371°C.(2500°F.and

soda/lime/silica glass melts at c.1482°C (2700°F.)
(Kulasiewicz, 1974).

These high temperatures would need a great deal of fuel, and the Anglo-Saxon glass workers probably built their furnaces near wooded areas, although few glass works have been recognised in archaeological sites. An indication of the amount of wood consumed is recorded at Knole (Kent) where in 1585/6 a fourpot glasshouse consumed 543 cords of wood in thirty-two weeks. (A cord is 128 cubic feet.) It has been calculated that this would clear about four acres of fifteen-year-old coppice wood each month. (Charleston, 1991). Although this is a considerable time after the Late Anglo-Saxon Period the technique was probably similar to that used by the Anglo-Saxons. It is of course possible that glass production was not carried out continuously throughout the year.

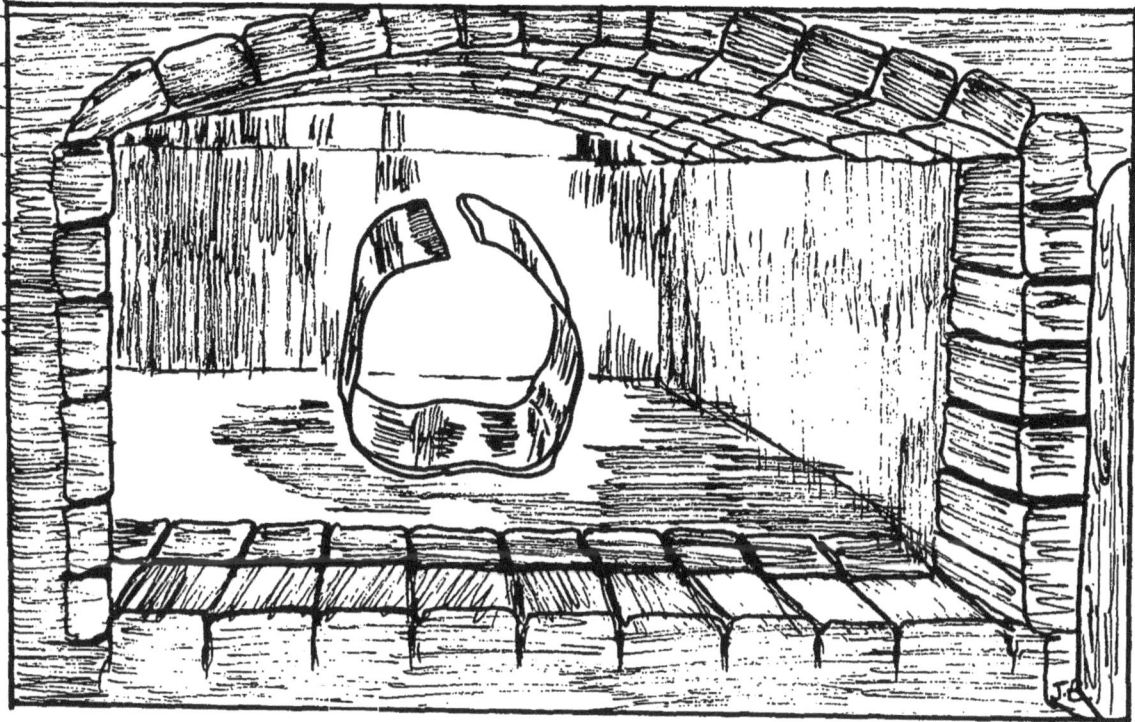

Figure7:1
Glass making

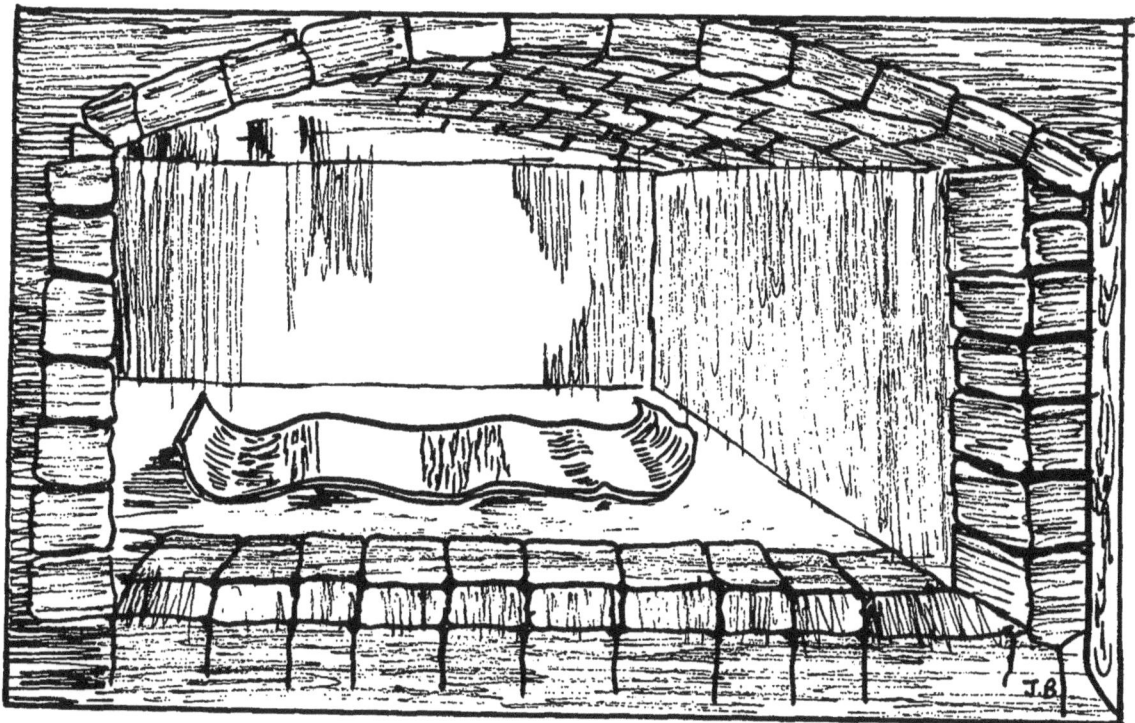

Smelting Lead and Iron in Anglo-Saxon Times

Bede wrote that 'Britain has also many veins of metals as copper, iron, lead and silver,' (Sherley-Price, 1955) Exports of lead are referred to in letters of 847-c.855 from Lupus of Ferrieres to King Æthelwulf asking for lead for a church roof.

> Eclesium in monasteno nostro, quod est mediterraneum, et Ferrarias appellatur ac Bethleem a conditore impositum nomen possidet, operine plumbo molimur post Deum In honore beati Petri et omnium icterorum apostolorum consecratum. (Episolarum Tomus Karolini, 1925)

> We are striving to cover with lead the church in our monastery which lies inland, and is called Ferrieres, and bears also the name of Bethlehem, bestowed on it by its founder, and is consecrated in honour, after God, of the blessed Peter and all the other Apostles. (Dorothy Whitelock, 1955.)

Plate 7:1 shows a cleft in the hillside, known as Odin's Mine, from which lead has been mined. *Domesday: Derbyshire* records a lead render from Hope, a village 2 miles away, with two other manors in the Peak District.

> Hæc tria Manoria reddebant. T.R.E. xxx lib et V sesta rios mell *et* dimidiu' et v. plaustrates plubi de. L. tabulis.

> Before 1066 these three manors paid £30, 5½ sesters of honey and 5 wagon loads of 50 lead sheets. (Morgan, 1978).

Although the place-name tempts one to associate this mine with lead production in Anglo-Saxon times there must be an element of caution. Cavers and climbers have a penchant for giving fanciful names to features, and the word 'sitch' appears in the *English Dialect Dictionary*, which indicates that it was in use after 1704. (Joseph Wright, 1904) The word 'grave' instead of 'mine' would be more indicative of a Saxon origin; cf. Grimes Graves, Norfolk, (see above).

Technical Background Most common metals can be extracted by roasting the ore with charcoal with a good blast of air. The melting temperatures are as follows:

tin	232°C	449°F
lead	327°C	620°F
silver	960°C	1760°F
gold	1063°C	1944°F
copper	1083°C	1987°F
iron	1535°C	2795°F (Hicks, 1963)

Melting temperatures should not be confused with smelting temperatures. Lead, for instance, occurs in the form of its sulphide, known as galena. The traditional method of smelting lead consists of roasting the ore, which removes the sulphur, leaving lead oxide. This is then heated with charcoal to produce lead.

The temperature at which lead is smelted in modern furnaces is about 1150°C because of the energy required to break down the chemical bonding between the metal and sulphur, (T.Davey, 1979). It is however possible to produce lead at the lower temperature reached in a wood fired brick built brazier. (R.F.Tylecote & J.F.Merkel, 1985).

Boltz in his article on 'Smelting' considers that cast iron was never made in any primitive furnace (C.L.Boltz, 1970) and Jane Geddes agrees that the high temperature was unlikely to be achieved until the end of the Middle Ages. (J.Geddes, 1991) Cast iron would be too brittle to be of much use because of the crystalline structure. The addition of 3% carbon lowers the melting point to 1150°C. This is also the temperature at which slag melts and can be poured off. (Geddes, 1991).

The iron, now called a bloom, was removed with tongs and beaten to remove the excess impurities. Wrought iron was therefore the usual product. This could be forged and welded and was more useful than cast iron would have been. (Geddes, 1991). In a modern experiment it was found that 16 lbs of charcoal were needed to produce one pound of iron. (Geddes, 1991)

The author has been unable to find documentary evidence for copper smelting in Anglo-Saxon England, possibly because copper deposits are mainly in Cumbria, Cornwall and Anglesey, which are areas from which few or no documents survive. The technique is illustrated by the old copper smelting furnaces at Lusaka, Zambia. The simplicity of the design indicates the ease with which copper can be extracted from its ore. Layers of charcoal and copper ore were placed in the furnace and ignited. The copper ore (mostly copper carbonate) was reduced, the copper melted and ran out of the hole at the bottom. The charcoal acted as a reducing agent and a source of heat. The same principle could be used to extract most metals from their ore. (Reducing is a chemical term for removing oxygen.)

Iron Smelting

The smelting of iron consumed a considerable amount of charcoal and the forging and smithing processes would have added to the amount. Evidence for the Anglo-Saxon iron industry is recorded in Domesday Book. Twenty-five places are mentioned, ten of which render blooms. Four have forges, five have smiths. There is one iron tithe and one iron mine. Three manufactured iron. One other mine (Orgrave) is deduced from a place-name only. (H. R. Schubert, 1957). These are summarised in the accompanying map and table. (Figures 7:2 and 7:3)

To find production figures it is necessary to extrapolate from the figures given for iron production in the Barony of Coupland, Cumberland, around the year 1240. It is likely that production methods did not change much between the Late Anglo-Saxon times and then, although mills were beginning to be introduced at the time of Domesday Book. (Schubert, 1957)

In one year 24 dozen of ore were used, and if the dozen as a measreure of ore remained constant at 12 hundredweights (as it was in the sixteenth and seventeenth centuries) then the amount of ore used was 14 tons 8cwt. (Schubert, 1957). If this yielded 17.5% (Schubert, 1957) of iron then 2 tons 14 cwt of iron would be produced.

It has already been shown that 16 lbs of charcoal are required to make one pound of iron (see above,), and that in charcoal making about one third of the wood is actually turned into charcoal (see above). From these figures we can calculate that 2 tons 14 cwt x 48 = 129 tons 12cwt of wood would be needed for a bloomery of this size. In order to relate this to trees the author cut down three ash trees, planted twenty years ago. The crown and branches were lopped off, as is current practice in charcoal burning. The average weight was twenty-eight pounds, giving eighty trees to make one ton. To provide enough charcoal for a bloomery for one year 129 x 80 = 11320 + 48 = 11,368 trees would be needed. We can therefore say that about 12,000 trees a year would be used. An acre is 4,840 square-yards, and as trees planted for coppicing on a fifteen to twenty year rotation are likely to be set at intervals of about four feet, and that rate of growth is not even, due to nutrients, light, aspect, and other variables such as soil depth, then a guesstimate of about an acre for 4, 000 trees would seem to be reasonable, making a requirement of 3 acres of coppice a year. On a twenty year coppice rotation about sixty acres of coppice would be required to provide a constant supply of fuel. This does not tell us any more than the possible amount which a bloomery the size of the one at Coupland may have used, but the sole aim of this exercise was to establish a concept of the magnitude of wood needed to support the iron smelting industry.

Plate 7:1

Odin's Mine, Castleton, Derbyshire. Map reference SK 148 832

This lead mine, now a cleft in the hillside, still has lead ore on the sides. A nearby stream runs in 'Odin's Sitch' (O.E.'sic' a water course) (Bosworth & Toller). Note the caution on the interpretation in the text.

Plate 7:2

Old copper smelting furnaces at Lusaka, Zambia

No. on map	County	Place Name	Designation of works	Working in 1066-1086
1	Sussex	Nr. E. Grinstead	ferraria (forge)	1086
2	Hampshire	Stratfield Turgis	ferraria	1086
3	Somerset	Alford	render of 8 blooms	1086
4		Bickenhall	render of 1 bloom	1066
5		Cricket St. Thomas	from every free man	
6		Glastonbury	8 smiths	1086
7		Lexworthy	2 mills rendering 2 blooms each	1086
8		Seaborough	1 bloom from every free man	1066
9		White Staunton	render of 4 blooms	1086
10	Gloucestershire	Alvington	render of 20 blooms	1086
11		Gloucester	render of manufactured iron	1066-1086
12		Pucklechurch	render of 90 blooms (massae ferri)	1086
13	Herefordshire	Hereford	6 smiths	1066-1086
14		Merchelai. Vicinity of Monmouth	render of I massa ferri. iron tithe	1066 before 1086
15	Northants	Corby	ferrariæ (forges)	industry probably
16		Gretton		destroyed in 1065
17		Norton	smiths	
18		Towcester	smiths	
19	Lincolnshire	Stow in Well	3 ferrariae	1086
20		Castle Bytham	3 fabricac ferri	1086
21		Little Bytham	fabrica ferri	1086
22	Derbyshire	Elvaston	I smith	1086
23	West Riding (Yorks.)	Hessle, part. of Wragby	6 ferrani ironworkers)	1086 1086
24	Cheshire	Rhuddlan	iron mine (now in Flintshire)	1086
25	Lancashire	Orgrave (Furness)	place name only	1086

Figure 7:2

Summary of iron workings in the late Anglo-Saxon Period. adapted from Schubert, *History of the British Iron and Steel Industry*

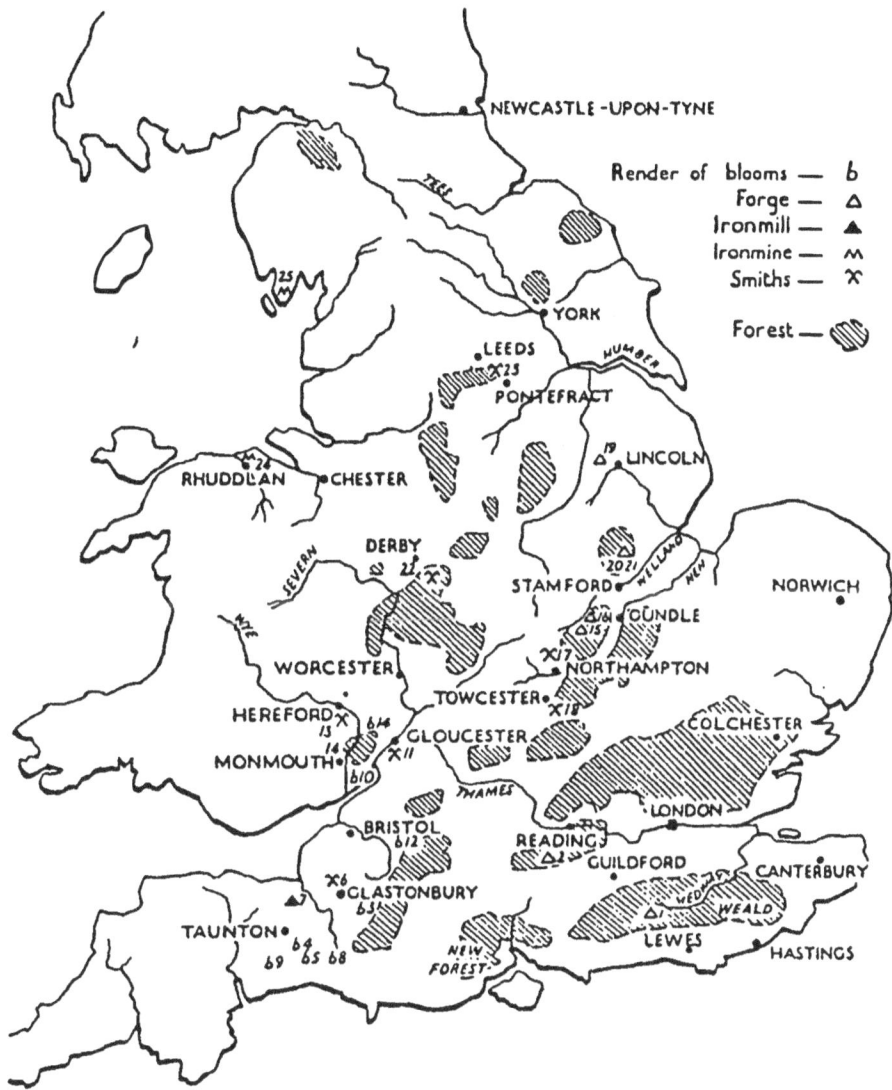

Figure 7:3

Iron working in the late Anglo-Saxon Period.

Conclusion

This study of the woodlands of the Late Anglo-Saxon Period has shown that they were a valuable resource and that consequently the uses, in the form of pannage, pasture, coppicing and the provision of timber were controlled by legislation, by grants of rights or privileges and management. Oliver Rackham outlined the criteria by which managed woodland could be recognised and the evidence provided by charters and the record of Domesday Book satisfy the criteria. There is evidence of named woods; the example of Elmstone Wood which was cited still survives as a place-name, four miles north-west of Prestbury, although the wood is no longer there.(Map reference SO 920 260.)

The land, and therefore the woodland, was owned, and the land owners are listed in Domesday Book. Charters record the transfer of leases and, as in the case of Elmstone Wood, the wood is named, or, as at Benson, Oxfordshire, the bounds of the wood belonging to the estate are described. Boundary banks still exist, and are described by Oliver Rackham in *The History of the Countryside*. Internal boundary banks, such as the one in Hockley Wood, Essex, (Plate 1:3) are still to be found.

The harvesting of a crop did not always consist of coppicing, but utilised the mast for pannage *silva porcorum* areas for pasture *silva ad pasturam* and specific uses such as wood for fencing *silva ad clausuram*. Coppicing *silva minuta* was an efficient method of maintaining a self renewing crop, providing firewood and building material.

The study of the coppice has shown that some traditional crafts, such as hurdle making, basketry, besom making and the growing of wooden forks, still continue The revival of the use of the pole lathe by bodgers (pole lathe turners) has been accompanied by the production of charcoal, used in the chemical industry and for barbecues.

Bark has always been in demand for the tanning of leather; five times as much bark as hides by weight, see above, but the use of oak woodlands for pannage is no longer practised in these days of factory farming.

Hurdles are still used, not in fish weirs, but in the traditional manner of the shepherd. The construction of barrels is a craft practised in the Late Anglo-Saxon Period, as the archaeological record and the Bayeux Tapestry show.

The use of the woodlands as a source of medicines is indicated by the *Leechdoms*, but the use of dyes from trees is one which we can only describe as a probability.

Leather occurs in the archaeological record as well as in Ælfric's *Colloquy* and the fact that the word alum (O.E. *efne*) occurs only once (as a gloss in a vocabulary) (above) suggests that it was not used in leather making. The alternative method of leather making. oiling the dried skins.

would not require tanning pits such as those found at Winchester. From this we can deduce that the method described earlier was the one in use in the Late Anglo-Saxon Period. We may also surmise that alum, although known, was used only with madder, where it is an essential mordant, and was probably a luxury item, and so limited to the richer classes.

The use of 'standard' trees in buildings has altered: they now appear as rafters and joists. The evidence from the door at Hadstock indicates the use of timber from a standard tree which was cut 'on the quarter', a technique which implies a good knowledge of the tree structure.

The manufacture of soap was a household task up to the last century, and the probability of it being so in the Late Anglo-Saxon Period has been explored by the author. Knowledge of soap in the Anglo-Saxon Period is indicated from Continental sources and the *Leechdoms*. The method uses lye (made from wood ash) and fat, two materials which must have been readily available. It is the author's opinion that soap was a familiar item in the home during the Late Anglo-Saxon Period. Wood ash was also a basic ingredient of glass. The change from the superior ingredient of natron to wood ash (which, incidentally, required a higher temperature to fuse the frit) opens up a sidelight on the extent of trade links with the Middle East at an earlier time in the Late Anglo-Saxon Period.

Charcoal was essential if the high temperatures of glass making and the smelting of metals were to be reached, and the by-product of tar during charcoal making is confirmed by the samples found in the Graveney boat. The smelting of lead and iron ore is indicated by the renders recorded in Domesday Book. The amount of wood consumed for glass making has been recorded and the implications should be considered. Four acres of coppicing consumed in thirty-two weeks is the equivalent of six acres a year. To provide this on a cycle of a fifteen year coppice rotation would mean ninety acres of coppice permanently devoted to glass production. (This is about the same area as eighty-five football pitches.)

Constant fires for salt evaporation would also make great demands, for the salt industry was well established, with salt roads networking the country.(D.H.Hill, 1981) Domestic consumption of fire wood would be constant, and the smaller industries, such as smelting and smithing, would consume vast amounts of fuel. It has been estimated that one pound of iron needed sixteen pounds of charcoal.(Geddes, 1991) This would need forty-eight pounds of wood. The amount of fuel needed for cooking is impossible to assess, but Ann Hagan, in her *Handbook of Anglo-Saxon Food Processing and Consumption* gives details for spit roasting a pig of 120 lb dead weight:

> The ideal fuel is 15 cwt of large oak logs, a foot long and thoroughly seasoned.--- on this 15 cwt of ash with a diameter of 4 - 5" cut in foot lengths (cut at least a month previously).(Ann Hagan, 1992)

One and a half tons of wood to roast a pig would make the operation expensive in time and labour, as well as fuel, and so be restricted to special occasions. To these scant figures we can add that the density of the woodlands will vary according to the soil, aspect, and the length of time between cutting the coppice. If we consider that an acre is 4,840 square yards, and that coppicing is not likely to have more than one tree for one square yard, and a greater area for trees which are allowed to grow on to become standards, then we may assume that about 4,000 trees an acre would be a good crop. These implications must make us aware of the careful management, and the great demands on the resources of the woodlands at the end of the Late Anglo-Saxon period. The paucity of remains reflects an economy based on organic materials. As we have seen, very little was wasted. In the past few years there has been an awakened interest in woodland crafts with the realisation that the woodlands are a renewable resource. The Anglo-Saxon civilisation was, in modern parlance, a 'green' society.

Appendices

1.The occurrence of the Latin phrase 'Cum onerat' in Domesday Book

The phrase in Domesday Book, *Cum onerat*, means 'when it bears' and is applied to woodlands to indicate an increased value when there is a large amount of mast, a phenomenon which occurs at intervals of three or four years

Warwickshire (interpreted in the Phillimore translation as 'when exploited.')

Sutton (Coldfield)	Chapter 1 section 7
Coleshill	Chapter 6 section 1
Roundshill	Chapter 6 section 4
Claverdon	Chapter 16 section 16
Preston (Bagot)	Chapter 16 section 18
Astley	Chapter 16 section 42
Smercote and in 'Sole'	Chapter 16 section 43
Bedworth	Chapter 16 section 44
Thorkell (in Coleshill Hundred)	Chapter 17 section 7
Ulverley	Chapter 42 section 1
Arley	Chapter 42 section 2
Fillongley	Chapter 44 section 10

Northamptonshire (interpreted in the Phillimore translation as 'when stocked.')

(Greens) Norton	Chapter 1 section 6
Oundle	Chapter 6 section 10a
Aldwincle	Chapter 6a section 27
Fotheringhay	Chapter 56 section 7

Oxfordshire (interpreted in the Phillimore translation as 'when stocked.')

Newington	Chapter 2 section 1
Witney	Chapter 3 section 1
Eynsham	Chapter 6 section 6
Stanton (Harcourt)	Chapter 7 section 3
Lewknor	Chapter 9 section 1

The same phenomenon is referred to by the term *'Si fructosa,* 'if it fruits', which occurs in Domesday Book under the entries:

Kent	
Wrotham	Chapter 2 section 10
Herefordshire	
Pembridge	Chapter 19 section 8

The opposite phenomenon is referred to by the term *infructosa,* 'barren', which occurs in Domesday Book under the entries:

Kent	
Canterbury	Chapter C 1
Dorset	
Nettlecombe (Exon]	Chapter 11 section 13
Renscombe	Chapter 11 section 16
Leicestershire	
Lubbesthorpe	Chapter 25 section 3

2.Dyes from Trees

Dyes which are obtainable from trees or tree parts are listed here as available to the Anglo-Saxons. It is not possible to determine which, if any, were actually used.

Red   Birch   *Betula alba*   Fresh inner bark

Blue   Elder   *Sambucus nigra*   Berries
(Saxon blue is made from extract of indigo, which was not made until 1740.) (E. Mairet, 1917)

Yellow

Privet	*Lingustruin vulgare*	Berries
Sloe	*Prunus communis*	Fruit
Birch	*Betula alba*	Leaves
Bramble	*Rubus Fructosus*	
Broom	*Sarothamnus scopius*	
Buckthorn	*Rhainnus frangula* and *R. catharica*	
	Berries and bark	
Crab apple	*Pyrus malus*	Fresh inner bark
Dyer's Greenwood	*Genista tinctora*	young shoots, leaves
Gorse	*Ulex europæus*	Bark, flowers and young shoots
Hornbeam	*Carpinus betulus*	Bark
Pear	*Pyrus* sp.	Leaves
Poplar	*Populus sp*	Leaves
Privet	*Lingustrum vulgare*	Leaves
Spindle tree	*Buorymus europæus*	
Wayfaring tree	*Viburnum lantana*	
Willow	*Salix spp*	Leaves.

Brown

Alder Alnus *glutinosa*   Bark
Birch *Betula alba*   Bark
Oak *Quercus* robur   Bark
Walnut *Juglans regia*   Root and green husks of fruit
Elder *Sambuscus nigra*   Leaves with alum
Privet *Lingustrum vulgare* Berries and leaves with alum

Purple
Elder *Sambuscus nigra* Berries give a violet shade, berries with alum and salt give a lilac shade. (Mairet, 1917)

Black
A true black is very difficult to achieve, but galls, from the oak, *Qercus infectoria* were used with iron

Theophilus tells how to make ink using hawthorn bark and iron sulphate. (Dodwell, 1961)

Wool and silk are receptive to dyes, but the other common fabric of Anglo-Saxon times, linen, is difficult to dye and is normally bleached. (Mairet, 1917)

From this list we can see that trees were potential sources of dyes, many of them substantive, and most of them from fairly common trees. It must be remembered that this list was published in 1917, and shows what the resources were, but there is no evidence which of these dyes were used in the Late Anglo-Saxon Period.

3. Making Lye and Soap

The first step in making soap is the preparation of lye. The traditional method is to bore holes in the bottom of an old barrel and to put a layer of gravel covered with straw for drainage. The barrel is then filled with wood ash and rain water is poured on. Eventually lye trickles out of the bottom. (F.Sherwood Taylor, 1957, and J.Seymour, 1984)

Historically rainwater was acidic because of dissolved carbon-dioxide, which gave carbonic acid. The present day air pollution contains sulphur dioxide, which dissolves in rainwater to form sulphurous and sulphuric acids. Because of this de-ionised water was used in the following experiment as curds of sulphites and sulphates may have formed in the soap-making process.

Experiment

Twenty litres of de-ionised water were mixed with wood ash from a mixture of hard and soft woods of unknown varieties, and left to stand for eight weeks. The concentration was such that a sludge formed at the bottom of the containers, so that the resulting lye would be as strong as possible. The liquid was a rich brown colour and this was then strained through nylon tights to remove particulate matter and then filtered through four thicknesses of cotton cloth, but no more impurities were removed. The lye was then reduced in volume by heating in an enamel bowl. Enamel was chosen because aluminium reacts with alkaline solutions, and iron, the metal probably used in Anglo-Saxon times, is not so readily available. Earthenware pots may have been used by the Anglo-Saxons but their porous nature makes the process less practical in a modern kitchen.

When the concentration was such that an egg would float in it the volume was about one third of the original. (Sherwood Taylor, 1957) Some water had been lost to the sludge and the resulting volume was five litres. At this concentration crystals of potassium carbonate were precipitated. The lye was later tested and found to have a pH of 13·3502.

The practice in medieval times was to sharpen the lye with quicklime, and Sherwood Taylor believes this to have been the method used from antiquity. (Sherwood Taylor, (1957), and Seymour (1984)) As quicklime is not readily available it was decided to make it. This was done by heating crushed limestone bought at a garden centre. It was placed in an iron paint kettle and heated to red heat on an open fire. This was done in small batches, each batch being kept at red heat for at least half an hour, by which time the limestone was changed to quicklime.

$$CaCO_3 = CaO + CO_2$$

This was not apparent until the limestone had cooled and it was then possible to see a fine white powder among the grey granules. It was clear that not all the limestone had byslaking with distilled water, leaving the unchanged limestone as a sludge and the slaked lime in solution.

The fifteenth century recipe which follows recommends that one part of quicklime be mixed with two parts of lye. (Sherwood Taylor, 1957). It was important that comparable strengths be mixed, and so the lime was kept as a saturated solution, and the specific gravity adjusted by adding distilled water until both solutions were the same. This was done by preparing a saturated solution of salt and weighting a stick so that it floated upright. A mark showed the level and this simple hydrometer enabled the lye and lime to be of the same densities. Two litres of lye were then mixed with one litre of lime water. (Sherwood Taylor, 1957) The proportions of fat to lye are traditionally about one litre of lye to one kilogram of fat. In this case lard was used and boiled for three hours. (Seymour, 1984) F.Sherwood Taylor quotes an early recipe from *The Secrets of Master Alexis of Piedmont*, written about 1547, and part is reproduced here as it was a guide to the foregoing experiment. (Sherwood Taylor, 1957)

> Take strong lye, with two parts of the ashes of the wood of the tree called in Latin *cerrus*. which is a kind of tree like to a poplar, having a straight long stem bearing a kind of mast, rough without like a Chestin, and one part of quicklime, and make it so strong that it may bear a new laid egg swimming between two waters. Take eight potfuls of this lye very hot, a potful of deer's grease or suet well strained clean: mingle them and set them on the fire, but see that they seeth not.

This recipe continued with advice about stirring it and letting the sun warm it, and then adding rose water to the paste and allowing it to set.

The strength of the quicklime was found to be of a pH 12·6812, and the mixture of lye and quicklime pH 13· 4082. As the pH scale ends at 14·000 it can be appreciated that the Anglo-Saxon lye makers were dealing with strong corrosive chemicals.

pH of lye

Indicator : Methyl Orange

25cc lye titrated against 2·8cc HCl at 2M

$$KOH + HCl = KCl + H_2O$$

2·8cc of HCl = $2,8 \times 10^{-3} \times 2$

$1dm^3$ KOH $= \dfrac{2·8 \times 10^{-3} \times 2 \times 10^3}{25}$

$= \dfrac{5·6}{25}$ Moles $= 22·4 = 0,224$ Moles

p(OH) = $-\log 10\ 0,224$

$= -1,3502 = +1\ -0,3502 = 0,6458$

pH $= 14,00 - 0,6458 = 13,3502$

Indicator : Methyl Orange

Ca $(OH)_2$ saturated solution 25 cc Ca $(OH)_2$ titrated against

0,6 cc HCl at 2M

$$Ca(OH)_2 + 2HCl \rightarrow CaCl_2 + H_2O$$

25 cc ?M Ca$(OH)_2 = 0,6$cc 2M HCl

1 litre of 2M HCl contains 2M

0,6 litre of 2MHCl contains $\dfrac{2 \times 0·6M = 1,2 \times 10^{-3}}{1000}$ Moles

$1dm3$ of Ca$(OH)_2$ contains $\dfrac{0\ 6 \times 10^{-3} \times 10^3}{25}$

$= \dfrac{0·6}{25}$ Moles $= \dfrac{2·4}{100} = 0,024$Moles

p(OH) = $-\log_{10}[OH]$

$= -\log_{10} [0,048]$

$= 2,6812 = +2 + (-0,6812 = +1,3188$

Therefore pH = $14,000 - 1·3188 = 12·6812$

pH measurements are on a logarithmic scale so that 13,00 is 10 times stronger than 12:00.

pH of the mixture of lye and quicklime

Indicator: Methyl Orange

K_2CO_3 + Ca$(OH)_2$ saturated solution

25cc titrated against 3,2cc of HCl at 2M

$$(Ca(OH)_2)(KOH) + 3HCl = (CaCl)(KCl) + 3H_2O$$

3,2cc of 2M HCL = $1 \times 2 \times 3,2 \times 10^{-3}$

25cc lye + quicklime = $2 \times 3,2 \times 10-3$

$1dm. \ldots = \dfrac{2 \times 3·2 \times 10^{-3} \times 10^{\cdot 3}}{25}$

$\dfrac{6,4}{25} = 0·256M$

p(OH) = $-\log_{-10} 0,256 = -1,4082$

$= +1 - 0,4082 = 0,5918$

therefore pH = $14 - 0·5918 = 13·4082$

Bibliography

M.Abbott, *Green Woodwork* Lewes, 1989 (reprint 1992).

S.0. Addy, *Evolution of the English House*, London, 1898 (revised. 1933),

P.D.Ancona and E.Aeschlimann, *The Art of Illumination*, London, 1969

Anonymous, *Builder*, 7, pp.22, 45, 115, London, 1849.

Anonymous, *Greensted Church Guidebook*, Ongar, undated

J.Arnold, *All Made By Hand*, London, 1970 (reprint 1975).

T.Arnold (ed.) *Memorials of St. Edmund's Abbey*, 1, Rolls Series 96, London, 1870.

J Backhouse, D. H. Turner, L. Webster, *The Golden Age* of *Anglo-Saxon Art*, London, 1984.

K.C.Bailey, *The Elder Pliny's Chapters on Chemical Subjects* II, London, 1929-32.

F. Barlow, (ed. & tr.), *Vita Ædwardi Regis: attributed* to a *monk of St. Bertin*, London, 1962.

J. Bately,(ed.) *The Old English Orosius*, E. E. T. S., S.S. 6, Oxford,1980.

M.Biddle, 'Report on Excavations at Winchester 1962-63', *Antiquaries Journal*, 44, 1964. pp.188-219.

W.De Gray Birch (ed.), *Cartularium Saxonicum: a collection of charters relating* to *Anglo-Saxon History* published in four volumes, London, 1885-1899.

C. L. Boltz, 'Iron and Steel' 3.1 'History' in J.H.Fearon (ed.), *Materials and Technology* III, London, 1970, pp.123-233.

W.Bonser, *The Medical Background of Anglo-Saxon England*, London, 1963.

J.Bosworth and T. N. Toller, *An Anglo-Saxon Dictionary Based on the Manuscript Collection of the Late Joseph Bosworth*, edited and enlarged by T.Northcote Toller , London, 1838-1921 and *An Anglo-Saxon Dictionary Based on the Manuscript Collection of Joseph Bosworth, Supplement by T.Northcote Toller with revised and enlarged Addenda by Alistair Campbell*, Oxford, 1921,reprinted 1972.

L.Cantor, *The Medieval Parks of England - A Gazetteer*, Loughborough, 1983.

M.O.H.Carver, 'Contemporary Artefacts Illustrated in Late Anglo-Saxon Manuscripts.' *Archaeologia*, 108, 1986. pp.117-145.

Hakon Christie, Olaf Olsen and H.M.Taylor 'Greensted Church' *Antiquaries Journal*, 59, 1979, pp.92-112.

R.J.Charleston 'Vessel Glass' in J.Blair and N.Ramsey, (eds.), *English Medieval Industries*, London, 1991, pp.237-264.

John Cherry 'Leather' in J.Blair and N.Ramsey (eds.), *English Medieval Industries*, London, 1991, pp.295-318.

T.O.Cockayne,(ed.) *Leechdoms, Wortcunning and Starcraft of Early England*, published in 3 volumes, Rolls Series 35, London,1864-1866.

T O. Cockayne,(ed.) *The Shrine: A Collection Of Occasional Papers On Dry Subjects*, Published in 13 parts, London, 1864-1870.

Thomas A.Cornicelli, *King Alfred's Version of St. Augustine's Soliloquies*, Harvard, 1969.

J.K.Crellin and J.Philpott, *Herbal Medicine Past and Present*, II, Durham, 1990.

Grace Crowfoot, 'The Braids', in C.F.Battiscombe (ed.) *The Relics Of Saint Cuthbert*, Oxford, 1956,pp.433-463.

Richard Darrah 'Working Unseasoned Wood' in Sean McGrail (ed.), *Woodworking Techniques Before A.D. 1500'*, BAR(SS) 129 (1982:). pp.219-229.

N.Davey, 'A Pre-Conquest Church and Baptistry at Potterne' *Wiltshire Archaeological Magazine 59*, 1964, pp.116-123.

T.Davey, 'The Physical Chemistry of Lead Refining' in J.M.Cigan, T.S.Mackey and T.J.O'Keefe *Lead-Zinc-Tin '80 TMS-AIMC World Symposium on Metallurgy and Environmental Control*, Warrendale, Pennsylvania, 1979, pp.477-507.

E.T.De Wald, *The Illustrations of the Utrecht Psalter* Princeton, undated.

B.K.Davison and R.Mackay, 'Medieval Britain in 1970' *Medieval Archaeology* 15, 1971, pp.130-131.

C.R.Dodwell, *Theophilus De Diversis Artibus*, London, 1961,

D.C.Douglas (ed.), *Feudal Documents from the Abbey of St. Edmunds*, Oxford, 1932.

H.L. Edlin, *Woodland Crafts of Britain*, Newton Abbot, 1949, reprinted 1973.

H.L. Edlin, *The Forester's Handbook*, London, 1953.

John Emsley, 'Plant A Tree For Chemistry', *New Scientist* 116, 8 October, 1987, pp.39-42.

Margaret L.Faull and Marie Stinton, (eds.), *Domesday Book 30:Yorkshire* 1 and 2, Chichester, 1986.

V.Fenwick, 'The Graveney Boat - a Tenth Century Find From Kent' *British Archaeological Reports* (British Series) 53 (1978).

R.J.Forbes, *Studies in Ancient Technology* IV, Leiden, 1964.

H.W.Fowler and F.G.Fowler, *The Concise Oxford Dictionary of Current English*, fourth edition revised by E.McIntosh, Oxford,1950.

G.N.Garmonsway,(ed.) *Ælfric's Colloquy*, London, 1939, revised edition, Exeter, 1991.

Jane Geddes, 'Construction of Medieval Doors' in Sean McGrail (ed.), *Woodworking Techniques Before 1500* BAR (SS) 129 (1982) pp.313-325.

Jane Geddes, 'Iron' in J.Blair and N. Ramsey (eds.), *English Medieval Industries*, London, 1991 pp.167-188.

P.Grierson, 'Relations between England and Flanders before the Norman Conquest', *Transactions of the Royal Historical Society*, 4th series, vol.23, 1941, pp.71-112,

Su Grierson, *Dyeing and Dyestuffs*, Aylesbury, 1989

Ann Hagen, *A Handbook of Anglo-Saxon Food Processing and Consumption*, Pinner, Middlesex, 1992, reprint 1993.

R.A.Hall, 'Tenth Century Woodworking in Coppergate, York', in Sean McGrail (ed.), *Woodworking Techniques Before AD. 1500* BAR (SS) 129 (1982 pp.231-244.)

N.E. S. A. Hamilton (ed.), *Willelmi Malmesbiriensis Monachi De Gestis Pontificum Anglorum* Rolls Series 52, London, 1870.

D.B.Harden, 'Glass and Glazes' in C.Singer, E.J.Holmyard, A.R. Hall and Trevor I Williams (eds.), *A History of Technology 2, The Mediterranean Civilisations and the Middle Ages C. 700 B.C, to A. D. 1500*, Oxford, 1956, pp.311-346.

D.B.Harden, 'Domestic Window Glass, Roman, Saxon, Medieval', in F.M. Jope (ed.) *Studies in Building History*, London, 1961, pp.39-63.

D.B.Harden, 'Ancient Glass III Post-Roman', *Archaeological Journal* 127, 1971, pp.78-117

C.Hardwick (ed.), *St. Matthew*, from work by J. Kemble. Cambridge, 1958.

H.L. Hargrove (ed.), *King Alfred's Old English Version of St' Augustine's Soliloquies*, New York, 1902.

C.A.Hewett, 'Tool Marks on Surviving Works from the Saxon, Norman, and later Medieval period' in Sean McGrail (ed.), *Woodworking Techniques Before 1500* BAR (ss) 129 (1982) pp.339-348.

C.A. Hewett, *Church Carpentry*, London, 1982.

J.Hicks, *Comprehensive Chemistry*, London, 1963.

D.H.Hill, An Atlas of Anglo-Saxon England, Oxford, 1981

J.Hill, The Complete Practical Book Of Country Crafts, Newton Abbott 1979.

Della Hooke, 'The Droitwich Salt Industry', *Anglo-Saxon Studies in Archaeology and History* 2. BAR(BS) 92 (1981), pp.123-169..

Della Hooke, *Anglo-Saxon Landscapes of the West Midlands* BAR (BS) 95 (1931).

W.Horn and B.Born, *The Plan of St.Gall* I and II, California, 1979.

G.W.Hoskins, *The Making of the English Landscape*, London, 1955, revised edition, 1977.

G.H.W.Humphreys, *The Manufacture of Sole and Other Heavy Leathers*, Oxford, 1966.

R.Hunt (ed.) *Ure's Dictionary of Arts, Manufactures and Mines* 5th edition, London, 1860.

M.R.James (introduction), *The Canterbury Psalter* London, 1935.

T.B. Johnston and J.Whillis (eds,), *Gray's Anatomy, Descriptive and Applied.* 31st edition, London, 1954 (1st edition 1858).

J.M. Kemble, (ed.), *Codex Diplomaticus Aevi Saxonici*, Published in six volumes, London, 1839-1848.

Wm. A. Keyser Jr., 'Steam Bending' in J. Kelsey (ed.), *Fine Woodworking - On Bending Wood*, Taunton, 1985, pp.16-24.

S.Keynes and M.Lapidge,(ed. & trs) *Alfred the Great*, Harmondsworth, 1983 (reprint 1988).

F.Kulasiewicz, *Glass Blowing* , New York, 1974.

E.T.Leeds, *Archaeology of Anglo-Saxon Settlements*, Oxford, 1913.

F.Liebermann, (ed.) *Die Oesetze der Angelsachsen* 1, Halle, 1903.

Charlton T.Lewis and Charles Short, *A Latin Dictionary founded on Andrews' edition of Freund's Latin Dictionary, Revised, Enlarged and in Great Part Rewritten.*, Oxford, 1879 (impression of 1966).

P.M.Losco-Bradley and C.R.Salisbury 'A Saxon and Norman Fish Weir at Colwick, Notts', in M. Aston (ed.) *Medieval Fish, Fisheries and Fishponds in England* BAR (BS) 182 (1988), pp.329-351.

H.R.Loyn and J.Percival (eds.), *The Reign of Charlemagne - Documents of Medieval History* II, London, 1975.

P.McGurk, D.N.Dumville, M.R.Godden and Ann Knock, (eds.), *An Eleventh Century Illustrated Miscellany* E. E. M. F. 21, Copenhagen,1983.

E Mairet, *A Book on Vegetable Dyes*, Hammersmith 1917.

M.Millet and S.James 'Excavations at Cowdrey's Down, Basingstoke,Hampshire, 1978-9', *Archaeological Journal*, 140, 1983, pp.151-279.

A.D. Mills, *A Dictionary of English Place Names*, Oxford, 1991.

E.G. Millar, (ed.), *Luttrell Psalter*, London, 1932.

Monumenta Germaniae Historica: Epistolarum Tomus VI, Karol mi Aevi IV, Berlin, 1925,

Philip Morgan, (ed.), *Domesday Book 26: Cheshire*, Chichester, 1978, edited from a draft translation prepared by Alexander Rumble.

Philip Morgan, (ed.), *Domesday Book 27; Derbyshire*, Chichester, 1978, edited from a draft translation prepared by Sara Wood.

Philip Morgan, (ed.), *Domesday Book 1: Kent*, Chichester, 1983, edited from a draft translation prepared by Veronica Sankaran.

Philip Morgan, (ed.), *Domesday Book 22: Leicestershire*, Chichester, 1979, edited from a draft translation prepared by Michael Griffin.

Carole A.Morris, 'Aspects of Anglo-Saxon and Scandinavian Lathe Turning', in Sean McGrail (ed.), *Woodworking Techniques Before 1500*, BAR(SS) 129 (1982) pp.245-261.

John Morris, (ed.), *Domesday Book 14: Oxfordshire*, Chichester, 1978, edited from a draft translation prepared by Clare Coldwell.

John Morris, (editor of text and translation), *Domesday Book 3:Surrey.* Chichester, 1975, edited from a draft translation prepared by Sara Wood.

John Morris, (editor of text and translation), *Domesday Book .2 Sussex.* Chichester. 1976. edited from a draft translation pre-

pared by Janet Mothersill.

John Morris, (ed.), *Domesday Book 23: Warwickshire*, Chichester, 1976, edited from a draft translation prepared by Judith Plaister.

Hilary Murray, *Viking and Early Medieval Buildings in Dublin'*, BARB (BS) 119 (1983),

J.A. H. Murray and H. Bradley, *New English Dictionary On Historical Principles*, Oxford, 1888-1928, W.A.Craigie and C.T.Onions, Supplement, 1933.

L.Musset, *Bayeux*, Bayeux, 1989.

A.Neuberger, *Technical Arts of the Ancients*, London, 1930.

Gale Owen-Crocker, *Dress in Anglo-Saxon England,*. Manchester, 1986.

O.Rackham, *Trees and Woodland of the British Landscape*, London,1976,

O.Rackham, 'Neolithic Woodland Management in the Somerset Levels' *Somerset Levels Papers* III, 1977, pp.65-72.

O.Rackham, *The History of the Countryside*, London, 1986.

C.A.Ralegh Radford , 'The Saxon *House'*,*Medieval Archaeology* 1, 1957, pp.27-38,

S. E.Rigold, 'The Anglian Cathedral of North Elmham', *Medieval Archaeology, 1* 1962-1963, pp. 67-108,

A.J.Robertson, (ed.), *Anglo-Saxon Charters*, Cambridge, 1939.

Stuart Robinson, *A History of Dyed Textiles*, London, 1969.

Alexander Rumble, (ed.), *Domesday Book 18: Cambridgeshire*, Chichester, 1981, edited from a draft translation by Jennifer Fellows and Simon Keynes.

L.F.Salzman, *Building in England Down to 1540*, Oxford, 1952.

P.H.Sawyer, *Anglo-Saxon Charters: an annotated list and Bibliography*, London, 1968.

H.R.Schubert, *History Of The British Iron and Steel Industry*, London, 1957.

J.Seymour, *The Forgotten Arts*, London, 1984.

L.Sherley-Price (ed.), *Bede: A History of the English Church and People*, Harmondsworth, 1955.

F.Sherwood Taylor, 'Pre-Scientific Industrial Chemistry' in C.Singer, E.J.Holmyard, A.R.Hall, and Trevor I Williams (eds.),*A History of Technology 2, The Mediteranean Civilisations and.the Middle Ages* C. 700B. C. to A. D. 1500, Oxford, 1956, pp.347-*373,*

F.Sherwood Taylor, *A History of Industrial Chemistry*, London, 1957.

C.Singer, *The Earliest Chemical Industry*, London, 1948.

W.W.Skeat (ed.), *St. Luke*, Cambridge, 1874,

A.H. Smith (ed.), *English Place Name Elements* 1,E.P.N.S. 25, Cambridge, 1956.

W.Steam, (ed.) Carl Linneaus *Species Plantarium, a facsimile of the first edition of 1753. vol 1,*, printed for the Ray Society, London, 1957.

W. H. Stevenson, *Asser's Life of King Alfred*, first edition 1904, new impression with an article by Dorothy Whitelock, Oxford,1959.

Sir F.M.Stenton, *Anglo-Saxon England*, third edition, Oxford, 1971.

M. Swanton, (ed) *Beowulf*, Manchester, 1978.

H.M.Taylor and J.Taylor, *Anglo-Saxon Architecture*, Cambridge, 1965.

E.Temple, *Anglo-Saxon Manuscripts 900-1066*, London, 1976.

Frank and Caroline Thorn, (eds.), *Domesday Book 7: Dorset*, Chichester, 1983, edited from a draft translation prepared by Margaret Newman.

Frank and Caroline Thorn, (eds.), *Domesday Book 17:Herefordshire*, Chichester, 1983, edited from a draft translation prepared by Veronica Sankaran.

Frank and Caroline Thorn, (eds.), *Domesday Book 21:Northamptonshire,* Chichester, 1979, edited from a draft translation prepared by Margaret Jones, Philip Morgan and Judith Plaister.

Frank and Caroline Thorn (eds.), *Domesday Book 25: Shropshire,* Chichester, 1986, edited from a draft translation prepared by Celia Parker.

B. Thorpe (ed.), *Codex Exoniensis: A Collection of Anglo-Saxon Poetry From A Manuscript in The Library of The Dean And Chapter Of Exeter,* London, 1842.

B. Thorpe, (ed.), *The Homilies Of The Anglo-Saxon Church 2 Ælfric,* London, 1846.

R.F.Tylecote and J.F.Merkel, 'Experimental Smelting Techniques: Achievements and Future' in P.T.Craddock and M.J.Hughes, (eds.), *Furnaces and Smelting Technology in Antiquity,* B.M. Occasional Paper no.48.London, 1985, pp.3-20.

P.F.Wallace R.Ó.Floinn,*Dublin 1000,* Dublin, 1988.

J.W.Waterer, 'Leather' in C.Singer, E.J.Holmyard, A.R.Hall and Trevor I. Williams, (eds.), *A History of Technology 2 The Mediterranean Civilisations and the Middle Ages, c. 700 B.C. to A.D.1500,* Oxford, 1955, pp. 147-190.

V.E.Watts, 'Medieval Fisheries in the Wear, Tyne and Tweed: the Placename Evidence' *Nomina 7 (1983)* pp.35-45.

Gaea and ShandorWeiss, *Growing and Using Healing Herbs,* Emmaus, Pennsylvania, 1985.

Dorothy Whitelock (ed.), *English Historical Documents 1 c.500-1042,* London, 1955.

K.P.Whitney, *The Jutish Forest,* London, 1976.

D.M.Wilson,'Crafts and Industry' in D.M.Wilson(ed.) *The Archaeology of Anglo-Saxon England,* Cambridge, 1976, pp. 253-281.

J.Wright (ed.),*English Dialect Dictionary:Being the complete vocabulary of all dialect words still in use, or known to have been in use during the last two hundred years.* London 1904.

Richard Paul Wülker (Editor and collator), T.Wright, *Anglo-Saxon and Old English Vocabularies* 2nd edition, Darmstadt, 1968.

www.ingramcontent.com/pod-product-compliance
Lightning Source LLC
Chambersburg PA
CBHW051307270326
41926CB00030B/4759

* 9 7 8 0 8 6 0 5 4 9 4 7 5 *